DARWIN'S FINCHES

David L. Lack, 1966
(credit: Ramsey & Muspratt, Oxford)

DARWIN'S FINCHES

BY

DAVID LACK

With Introduction and Notes by
LAURENE M. RATCLIFFE
AND
PETER T. BOAG
Edward Grey Institute of Field Ornithology
University of Oxford

CAMBRIDGE UNIVERSITY PRESS
Cambridge
London New York New Rochelle
Melbourne Sydney

CAMBRIDGE UNIVERSITY PRESS
Cambridge, New York, Melbourne, Madrid, Cape Town, Singapore,
São Paulo, Delhi, Dubai, Tokyo

Cambridge University Press
The Edinburgh Building, Cambridge CB2 8RU, UK

Published in the United States of America by Cambridge University Press, New York

www.cambridge.org
Information on this title: www.cambridge.org/9780521272421

First published 1947
Reissued with an introduction and notes 1983
Re-issued in this digitally printed version 2009

A catalogue record for this publication is available from the British Library

Library of Congress Catalogue Card Number: 82–19856

ISBN 978-0-521-25243-0 Hardback
ISBN 978-0-521-27242-1 Paperback

Additional resources for this publication at www.cambridge.org/9780521272421

DEDICATED

to the

STAFF OF THE CALIFORNIA ACADEMY OF SCIENCES

with grateful thanks for their hospitality in the summer of

1939

CONTENTS

PART II—INTERPRETATION

ILLUSTRATIONS

PLATES

Plates I–VIII are bound between pages xxxii *and* xxxiii

TEXT-FIGURES

PREFACE TO THE 1983 REISSUE

This volume reproduces David Lack's 1947 edition of *Darwin's Finches*, with facsimile pages of the original text, tables, and line illustrations. The colour plates have been rephotographed from the originals, and the black and white plates have been either rephotographed or replaced with similar but sharper versions. We have added a brief Introduction to the original text, giving an overview of the origins and impact of the book. At the end of our Introduction, we have also provided a Notes section, to guide the reader to particularly important points in the text, as well as to provide updated information in selected areas. At the end of our Notes, an extensive, modern bibliography is provided for additional reading. In our Introduction and Notes sections, we use the convention of citing references in the modern bibliography as '(Lack, 1969)', while references in Lack's original bibliography are cited as '(Lack, 1947 L.)'. In these new sections we occasionally refer to archival material, from the Alexander Library at the University of Oxford. The Lack archives are organized as a series of numbered files, identified by references of the form '(Lack archives, No. 99)'.

Many people helped make this volume possible. We thank C. M. Perrins, Mrs E. Lack, G. C. Varley, C. Elton, H. N. Southern, M. J. Kottler and E. Mayr for background material on David Lack. P. R. Grant, F. J. Sulloway, D. Schluter, T. D. Price, G. C. Varley and E. Mayr kindly reviewed the Introduction and Notes. A. Richford helped us in the Alexander Library, while the Cambridge University Library permitted us to reproduce the colour plates from their copy of *The Zoology Record of the Voyage of H.M.S. Beagle*. G. T. Corley Smith helped us to find photographs; we also thank R. Perry and F. Pölking for the use of their pictures. We were hosted by the Edward Grey Institute at the University of Oxford while preparing the book. We thank T. R. E. Southwood, C. M. Perrins and J. R. Krebs for making this stay possible. P. T. B. received financial support from a National Science and Engineering Council of Canada postdoctoral Fellowship. The project was suggested and ably managed by M. Walters of Cambridge University Press.

L.M.R. and P.T.B.

Oxford, July 1982

PREFACE TO THE 1961 EDITION

A condition of the 1961 reprint is that the text should not be changed. The reader may therefore be reminded that this text was completed in 1944 and that, in the interval, views on species-formation have advanced. In particular it was generally believed when I wrote the book that, in animals, nearly all of the differences between subspecies of the same species, and between closely related species in the same genus, were without adaptive significance. I therefore specified the only exceptions then known (see pp. 39, 65–66, 79–80, 88–90 and 143) and reviewed the various ideas as to how non-adaptive differences might have been evolved (see pp. 122–123). Sixteen years later, it is generally believed that all, or almost all, subspecific and specific differences are adaptive, a change of view which the present book may have helped to bring about. Hence it now seems probable that at least most of the seemingly non-adaptive differences in Darwin's finches (see pp. 43–44, 77–79 and 117) would, if more were known, prove to be adaptive.

Another cautionary note may be added. In Chapter XI, I followed the traditional view, still generally accepted, that the ancestor of Darwin's finches was a true finch. But if a warbler-like bird can be evolved from a finch, so can a finch-like bird from a warbler, and that most of the Geospizinae are finch-like might perhaps mean, not that their common ancestor was a finch, but that there were more vacant niches in the Galapagos for finches than warblers. Further, the only living species which possesses characters from which all the rest might be derived is the Cocos-finch *Pinaroloxias*, which in beak comes closest to the warbler-finch *Certhidea* but in plumage resembles *Geospiza*, especially the sharp-beaked ground-finch *G. difficilis*. I wish to imply here not that *Pinaroloxias* is necessarily ancestral, but that the question of which of Darwin's finches look most like their common ancestor is entirely open. Without fossil material to give perspective in time, it is impossible to be sure of which species are near the base and which near the top of the evolutionary tree.

D.L.

Oxford, December 1960

(Reproduced courtesy of
Harper Brothers, New York)

PREFACE

Just over a hundred years ago, in 1835, Charles Darwin collected some dull-looking finches in the Galapagos Islands. They proved to be a new group of birds and, together with the giant tortoises and other Galapagos animals, they started a train of thought which culminated in the *Origin of Species*, and shook the world.

In an English garden, finches of several kinds and sizes eat seeds and fruits, tits of various sizes examine the twigs and branches, warblers of different kinds take insects off the leaves, thrushes search the ground and woodpeckers climb the trunks. That all these birds have a common ancestry is generally agreed, but to determine the way in which this evolution has come about would require a detailed knowledge of conditions throughout Europe and Asia, both at the present time and in the past. That these birds to some extent share out the available food supply is also clear, though their ecological interrelationship is highly intricate.

A similar complicated picture is presented by the bird life of North America or of any other region of continental type. It is no accident that Darwin and many after him obtained their inspiration from the much simpler conditions provided by the land life of remote islands, and in this respect no group is more suitable for study than Darwin's finches of the Galapagos. These birds form a small and self-contained group, but are sufficiently diverse to provide a parallel with evolution elsewhere, and every stage is represented from that of trivial differences between island forms to that of genera with striking adaptations to new and quite unfinchlike ways of life. The object of this book is to trace these successive stages, and many of the conclusions apply, I believe, much more widely than to the Galapagos finches.

This is a work of natural history, based on a study of living birds in the Galapagos and of dead specimens in museums. The evidence is circumstantial, not experimental, so that theories must be presented cautiously. They should not, however, be excluded. The Galapagos naturalist may lose himself in misty highlands with densely tangled trees, but in the open lowlands he may be parched in a lava desert. Likewise in this book I have tried to avoid not only intricate and nebulous theory, but, at the other extreme, an arid enumeration of bare facts.

An earlier and more specialized account of Darwin's finches is appearing as Occasional Paper XXI of the California Academy of Sciences. This was submitted in June, 1940, when I supposed I could not work further on the subject until the end of the war. But during 1943–4 I found time to write the present book, with the object of giving a broader treatment of evolutionary aspects. Unexpectedly, a reconsideration of the original material led to a marked change in viewpoint regarding competition between species and the beak differences between the finches, and the development of these points provides one of the main themes of the book. While I am astonished that what now seems obvious should have escaped notice four years earlier, there is a highly distinguished precedent, since even Charles Darwin appears not to have appreciated the evolutionary evidence provided by the finches, until several years after his return from the islands.

The many people who, in their writings or in person, have helped to make this book possible, are listed later, under References and Acknowledgements respectively. The debt is considerable. For further introduction, the reader should turn again to Darwin's description of his visit to the Galapagos in H.M.S. *Beagle*: 'The natural history of these islands is eminently curious, and well deserves attention.'

D. L.

London, August 1944

INTRODUCTION

Few books have had as great an impact on evolutionary ecology as David Lack's *Darwin's Finches*. Although it was first published almost forty years ago, it is still one of the most succinct and fascinating treatises ever written about Darwin's favourite subject – the origin of new species. It is especially appropriate that *Darwin's Finches* be reissued during the Darwin Centenary, when controversy about the mechanics of evolution continues to flourish. Were David Lack alive today, he would probably be gratified (though hardly surprised) that this book has withstood the test of time so well.

Lack's life-long passion for birds began at an early age in the marshes of rural Norfolk. At nine, he compiled his first life list, and at fifteen he began his first bird diary, having seen exactly 100 species. These schoolboy notebooks reveal the remarkable attention to detail in field observation that distinguished his later work. In 1929, Lack went up to Cambridge to read zoology, but apparently he did not enjoy his formal academic training, from biologists of the day such as Saunders, Carter and Salt. Later (Lack, 1973), he reminisced that his zoology course contained 'nothing about evolution, ecology, behaviour or genetics, and of course nothing about birds'! However, he enlivened his dry undergraduate programme by running the Cambridge Bird Club and going on expeditions to places like Greenland and St Kilda. At the end of his fourth year, he accepted a post as biology master at Dartington Hall in Devon. There he was able to indulge his love for field work and birding. Partly as a field exercise for his young pupils, he began a four-year investigation of territorial behaviour in the English robin. The study culminated in a classic text on passerine life history, *The Life of the Robin* (Lack, 1943). This stimulated him to attempt more comparative work.

The 1938–9 Galápagos expedition, on which this book is based, arose from Lack's desire 'to compare territorial behaviour in a group of related species' (Lack, 1973). Julian Huxley, Lack's long-time mentor, was instrumental in obtaining the necessary financial support for the trip, from both the Royal Society and the Zoological Society of London. After much delay, Lack and his party of five reached Wreck Bay on Chatham (now Puerto Bacquerizo Moreño on Isla San Cristóbal) in December 1938, at the start of the Galápagos rainy season. Then began a five-month struggle with rough terrain and intense heat, dysentery and unreliable transport, which strained the social relationships of the isolated team. A far cry from the modern comforts of the Charles Darwin Research Station, or even the comparative luxury of nineteenth-century collecting expeditions (when naturalists wined and dined aboard their ship each night!).

If the expedition was not a social success, it was a scientific success in most respects. It is a mark of Lack's skill in preparing for scientific expeditions that so much was accomplished in less than six months. Lack found that the finches were extremely easy to observe, although he had some trouble assigning certain individuals to the appropriate species at first. On Chatham (San Cristóbal), mornings were spent observing the feeding and breeding behaviour of the finches, while afternoons were spent building aviaries, in a futile attempt to get the different species to hybridize. Like most field biologists, Lack appears to have preferred to be on his own while collecting data. However, he (wisely) bowed to the rigours of the Galápagos climate, and rarely remained out after eleven a.m., mad dogs and Englishmen not withstanding.

At the end of January 1939, the expedition moved to Academy Bay on the south coast of Indefatigable (Santa Cruz). There they enjoyed the hospitality of the few European settlers, one family at a time since many of them did not speak to each other. The humid uplands of Santa Cruz provided much of the opportunity to observe tree finches (*Camarhynchus* spp.). Lack's field trips were enlivened by hunts for feral pig, which added variety to the otherwise monotonous diet of rice and vegetables. In January and early April, Lack was able to make short, one-day visits to Hood (Española) and Tower (Genovesa). On 4 April 1939, the main part of the breeding season over, the expedition sailed from Galápagos for Panama.

The departure was complicated by several cages full of ground finches (*Geospiza* spp.) which were destined for captive breeding studies in England. The birds began to die soon after leaving the islands. Lack spent several worrying days in Panama getting permission to send them instead to the California Academy of Sciences in San Francisco, and finding a ship that would take them. He himself profited from the unexpected side trip, working on the large collection of Geospizine skins at the Academy. At the outbreak of war in Europe, Lack departed for England, having written the first draft of the results of the expedition while in New York in September 1939. In 1940, after discussions with Huxley and a discouraging response from the editor of *Ibis* (Lack, 1941), Lack decided to publish the work as an Occasional Paper of the California Academy of Sciences (Lack, 1945 L.).

Owing to the vagaries of war and the shortage of editorial staff, the Academy did not publish the 160-page *The Galápagos Finches – a Study in Variation* until 1945, five years after Lack completed it. The book *Darwin's Finches* also had a long gestation. Lack completed the first draft in the winter of 1943–4, during a lull in his wartime duties as radar researcher. The book did not appear in print until 1947, and then it revealed quite a radical shift in emphasis from the earlier paper. It was Lack's new approach to old data which gave the book such impact, so it is worth looking at the genesis of Lack's fundamental change in interpretation in some detail.

In the California Academy paper, Lack followed traditional thinking at the time and concluded that while there were ecological differences between genera, most of the morphological differences between species or populations of Darwin's finches were non-adaptive, except perhaps as reproductive isolating mechanisms. There was no attempt to relate morphological variation below the generic level to variation in habitat, food supply or community structure. Most size differences among island populations of the same species were viewed as the result of chance phenomena such as genetic drift due to isolation. In other cases, Lack invoked hybridization to explain morphologically unusual populations such as those on Islas Daphne Major and Crossmans (Los Hermanos). In retrospect, the emphasis on non-adaptive explanation seems extraordinary, but Lack's view was widely accepted at the time, particularly in Britain. Huxley (1942 L.), in *Evolution: The Modern Synthesis*, agreed that most specific differences in the Geospizinae were non-adaptive, yet 42 pages earlier (p. 284), he described how different species of British finches have specialized feeding niches: 'They show how widespread is the tendency to ecobiotic and ecotypic differentiation – in other words to a specialized sharing out of the environmental habitat and ways of exploiting it among different related species.'

Lack himself must eventually have been struck with the ironic juxtaposition of his ideas with Huxley's in this book. In early 1943, he had begun plans for a text for sixth-form students, using the Galápagos finches to illustrate the principles of evolution. Instead, he now began to re-examine the finch data in terms of competition theory, as elucidated by the laboratory experiments of Gause (1934 L.) and Park, Gregg and Lutherman (1941). As early as 1939, Lack had read and 'immediately dismissed' Gause's ideas on competition, as laid out at the end of an article by Park (1939). On first reading Huxley (1942 L.), he had also failed to appreciate the idea that 'size differences...might have been evolved to reduce competition. I now rediscovered both ideas [Gause's and Huxley's] for myself' (Lack, 1973). He had several long discussions about interspecific competition with former Cambridge classmate and zoologist G. C. Varley, who was in the same small group working on radar during the war. Varley, whose interests were in insect ecology, recalls that many of these talks took place during an extended car tour of radar installations in southern England in 1943. We do not know the exact process by which Lack came to conclude that morphological variation in Darwin's finches is adaptive, but clearly, these conversations were influential (see also p. 6 in Lack, 1971). At some time in 1943, he became convinced that bill size was related to food type taken, and that differences in bill size among closely-related species reflected adaptations to different foraging niches. Gause's theory of competition explained why certain species of Darwin's finches with very similar foraging behaviours (e.g. *Geospiza conirostris* (large cactus finch) and *G. scandens* (cactus finch)) did not co-exist. It also explained why geographic replacements tended to occur among species with similar diet requirements. Chapter VI of *Darwin's Finches* outlines his changes of view in detail.

Lack devoted much of the rest of the book to formulating two important theses: that interspecific competition is a powerful force structuring animal communities, and that ecological isolation is as much a prerequisite for the co-existence of new species as is reproductive isolation. He incorporated both ideas into a model of the adaptive radiation of Darwin's finches through allopatric speciation. The model suggested that the ancestral colonizing flock of birds had differentiated into island forms, and that the divergence between these forms was completed when the populations re-established secondary contact. The cycle of differentiation was then repeated several times. As Grant (1981*a*) and Grant and Schluter (1983) explain, Lack overemphasized the importance of interspecific competition in promoting differentiation in sympatry because he could not see how populations would diverge very much ecologically in allopatry. Much later Lack (1969) acknowledged his mistake in assuming that different islands in the Galápagos were too similar for much local adaptation to occur. This point was first made by Bowman (1961), and later substantiated by Abbott, Abbott and Grant (1977). Despite the imbalance in Lack's initial theory, the concepts of interspecific competition and ecological isolation still form an important part of the framework of modern evolutionary ecology.

Lack argued from observation of existing pattern and not from any quantitative tests of theory. Perhaps it is not altogether surprising that his new ideas were not immediately accepted. At the Easter meeting of the British Ecological Society in 1944, Lack discussed the general importance of ecological isolation in avian speciation (Anon., 1944; Lack, 1944 L.). He was 'widely disbelieved, though...Charles Elton and I were on the same side' (Lack, 1973); Elton, it turned out, was also pioneering in the field of competition (Elton, 1946). But with his characteristic tenacity, Lack continued to investigate the regular patterns of morphology and feeding behaviour that he observed in bird communities. His papers on competition in shags and cormorants (Lack, 1945) and in birds of prey (Lack, 1946) are now considered classics. His ideas on ecological isolation became particularly well-known through his contribution to the volume *Genetics, Paleontology and Evolution* (Lack, 1949).

Understandably, the staff of the California Academy were also somewhat resistant to Lack's re-interpretation of the Galápagos work. The Occasional Paper on variation had just been published when Lack sent them the manuscript for *Darwin's Finches*. On 15 November 1945, Robert Miller, Director of the Academy, wrote:

'I have ventured to wonder whether the conclusions you drew as a consequence of your first contact with the data and while your field observations were fresh in your mind, may not be just as valid as those you have arrived at by further cogitation and study... You must not be too much disappointed if you find American workers citing your earlier conclusions in preference to your later ones' (Lack archives, No. 51).

On the same date, Robert Orr, who had just completed a laboratory study on the behaviour of the captive finches and was also disappointed by the new book, wrote: 'I rather regret to hear that you have changed your opinions in regard to so many matters relating to this group as I fear this will weaken rather than strengthen your position' (Lack archives, No. 51).

Against this negative opinion Lack fortunately received the positive support of people like Huxley, Mayr, Southern, Varley and Thorpe, several of whose works had influenced his own writing (e.g. Mayr, 1942 L.). In his comments on the proofs of *Darwin's Finches*, Huxley (14 March 1946) urged Lack to sub-title the book 'A Study in Evolution, or something similar, because otherwise people will not realize what an important contribution you have made to speciation in general'. Lack was disinclined to do this, and wrote to F. H. Kendon, his editor at CUP, on 19 March 1946 saying so 'because I dislike spoon-feeding the reader. Huxley probably feels that the zoological reader may think the book a popular work on ornithology and therefore not trouble to see it, whereas it is a book for biologists rather than ornithologists'. But Kendon prevailed, replying to Lack that 'it will certainly go on the jacket in some way...It is impossible to spoon-feed the reader with the label on the bottle...And won't those poor bird-lovers who buy the book under a misapprehension have a just complaint?' (Lack archives, No. 51). (Curiously, the sub-title was actually omitted from the 1947 edition, although it appears on the 1961 reprint.)

By the time the book actually appeared in 1947, Lack had switched his attention from competition and speciation to the problems of population regulation and avian reproductive patterns. Possibly this explains why the book, although enthusiastically received, seemed to lie in a vacuum for ten years. Not until the late 1950s and early 1960s did its main ideas begin to influence community ecologists. Ironically, the major stimulus was to workers in North America rather than in Britain, with ecologists such as Hutchinson (1951, 1957), MacArthur (1958, 1972), and their colleagues (Pianka, 1974; Cody and Diamond, 1975) taking up where Lack had left off.

There is no question about Lack's importance in bringing competition theory into animal ecology (Jackson, 1981), although it is probably too soon to assess the long-term impact that competition theory has had on the development of ecological thought. Few modern biologists deny that competition plays a role in the structuring of communities, but there is increasing concern that the uncritical acceptance of competition as a guiding principle in community ecology may hamper the search for alternative explanations of ecological patterns. Grant (1977) points out that Lack himself (1976) was guilty of pushing competition too far on occasion. This is not surprising, given that Lack spent much of his professional career collecting vast amounts of inferential evidence for competitive processes in nature. His forte was the consummate ability to watch, detect, synthesize, and describe patterns in the avian world. He was less interested in deducing testable predictions from his novel hypotheses, and methodically confirming or rejecting those hypotheses with fresh, quantitative data. Plant ecologists had applied experimental techniques to competition questions much earlier, but animal ecologists did not do so until well into the Hutchinson–MacArthur era (Jackson, 1981). Lack's shortcomings in this area may reflect a trade-off between generality and precision, necessary for progress in a developing science (Levins, 1968).

Darwin's Finches has remained a classic for two reasons – for its extremely accurate descriptions of geospizine ecology and behaviour (which even present-day researchers will appreciate), and for its original, cogent formulation of the adaptive radiation model. But it should be remembered that the formulation itself was not based on any strong quantitative data base. For example, Lack invoked interspecific competition for food as a major diversifying force in the radiation. However, he did not identify the range of food plants important for the finches, nor did he measure diet overlap between species. Similarly, he theorized that speciation was achieved by the reinforcement of reproductive isolation among sympatric populations, but did not clearly establish what those isolating mechanisms were. Modern research on Darwin's finches has been devoted largely to filling in these gaps with a view to testing competition and speciation hypotheses.

While Lack emphasized interactions among species, current workers have sought to characterize more adequately the interactions between finches and their environments. Bowman (1961) made an important start by describing how different species of Darwin's finches actually partition the available food resources. More recently, a great deal of effort has been devoted to measuring both the food handling abilities of ground finches (*Geospiza* spp.), and the temporal and spatial variation in food supplies on different islands (Abbott, Abbott and Grant, 1975; Grant *et al.*, 1976; Abbott *et al.*, 1977; Smith *et al.*, 1978; Grant and Grant, 1980*a*, *b*; Grant, 1981*b*; Boag, 1981; Schluter, 1982*a*, *b*; Schluter and Grant, 1982). In addition, two long-term studies with marked birds on Daphne Major and Genovesa islands are yielding some of the first baseline information on finch breeding behaviour and population dynamics (Grant and Grant, 1980*a*; Boag, 1981). Recent investigations of finch vocal behaviour (Bowman, 1979, 1983; Ratcliffe, 1981), a topic almost ignored by Lack because he had no recording equipment, reveal that song variability parallels and even exceeds morphological variability. The interaction of vocal and morphological signals in species recognition and mate selection has been the focus of work by Grant and Grant (1979) and Ratcliffe (1981).

It is difficult to summarize in a few lines the new insights provided by these studies. Basically, strong support for the competition hypothesis has emerged (Abbott *et al.*, 1977; Schluter and Grant, 1982; Grant, 1983; Grant and Schluter, 1983), despite a certain amount of controversy (see Grant and Abbott, 1980). There is now good evidence that much of the morphological variation in ground finches is inherited (Boag and Grant, 1978; Boag, 1983), and we know that selection, mediated through rainfall and its effects on food supply, can have a dramatic effect on finch phenotypes (Boag and Grant, 1981). Studies on the finch fossil record (Steadman, 1981) and on genetic variation in finch populations (Yang and Patton, 1981) are also helping to clarify and enlarge Lack's model of the adaptive radiation of the finches.

Future research on the finches needs to be aimed at understanding how intraspecific variation becomes translated into interspecific variation (e.g. Grant, 1981*a*, *c*; Grant and Price, 1981). We also need to find out whether similar processes of local adaptation have been significant in the radiation of the hitherto neglected tree finches (*Camarhynchus* spp.) and the warbler finch (*Certhidea*). In *Darwin's Finches*, as in so many other of his works, Lack provided us with the blueprint for study.

NOTES

CHAPTER I

pp. 2–5. Lack's bleak account of the islands was in part a response to the difficulties his group encountered in basic living and field work. But his private field journal also includes many references to 'splendid days of birding' and 'madrigal singing', which would seem to belie this sombre introduction. Lack used the English names for the islands for historical continuity. The Ecuadorian names (see Lack's Table 1 on p. 7) are used today. Culpepper and Wenman are called Darwin and Wolf, while Santiago and Floreana are also called San Salvador and Santa María, respectively. Lack gives the general impression that the islands are similar to each other in climate and physiognomy. This is mistaken, as floristic (Bowman, 1961; Abbott *et al.*, 1977) and climatic (Grant and Boag, 1980) analyses show. The most comprehensive guide to Galápagos plants (Wiggins and Porter, 1971) is continually being revised as more and more islands are intensively sampled. Factors contributing to differences in island vegetation include variation in island area, altitude, aspect, isolation and rainfall (Abbott *et al.*, 1977). The amount of precipitation any one island receives in a year is particularly unpredictable (Grant and Boag, 1980). Good descriptions of the altitudinal zonation of plants in Galápagos can be found in Wiggins and Porter (1971) and Harris (1974). Some upper transition/humid zone forest is found on Fernandina (D. Schluter, personal communication), in addition to the other locations mentioned here. There is also a small area of open country above the humid zone on Isla Pinta.

p. 5. The continuing encroachment of feral animals endangers several Galápagos species, including the land iguana (*Conolophus* spp.), the Galápagos tortoise (*Geochelone*) and the Hawaiian petrel (*Pterodroma phaeopygia*) (Duffy, 1981*a*, *b*). Conservation measures are being implemented by the National Park Service of Galápagos and the Charles Darwin Research Station (CDRS). Status reports on species at risk can be found in issues of Noticias de Galápagos, the report of the Charles Darwin Foundation. Annual reports of the CDRS provide additional summaries of recent work in Galápagos.

pp. 6–8. The human population of the Galápagos has grown to approximately 5000, although there have been no new settlements established since Lack's visit. The economy is based on tourism, farming and fishing. The Charles Darwin Foundation was established in the late 1950s with the aid of UNESCO, IUCN and the government of Ecuador. The Foundation was instrumental in helping to establish the Galápagos as a National Park of Ecuador in the early 1960s. There has been a dramatic rise in tourism since 1973, when regular flights from the South American mainland were initiated. Currently, the Park Service tries to limit the number of tourists to approximately 10000 per year but this is frequently exceeded. General accounts of the history of the islands can be found in Eibl-Eibesfeldt (1960), Thornton (1971) and Moore (1980).

CHAPTER II

pp. 13–18. Sulloway (1982*a*: 45–6) has pointed out that the design of Lack's book and his popularization of the name 'Darwin's finches' helped to perpetuate the legend that the geospizines provided the major stimulus to Darwin's theory of evolution, when in fact they were not even mentioned in the *Origin of Species* (Darwin, 1859 L.). Sulloway (1982*a*, *b*) has also argued convincingly for the renaming of *G. difficilis* as *G. nebulosa*. Apart from this, Lack's system of 14 species in four genera is still generally accepted, as are the English names he coined for the finches. The most common variation is the elevation of *C. crassirostris* and *C. pallidus* to distinct genera or sub-genera, *Platyspiza* and *Cactospiza* respectively (e.g. Harris, 1974). The nomenclature of closely related groups of island birds is inevitably somewhat arbitrary, but current evidence provides no obvious justification for the extra genera (e.g. Yang and Patton, 1981), and recent check-lists do not use them (Paynter, 1970; Howard and Moore, 1980; Walters, 1980). Steadman (1982) suggests that Darwin's finches are so closely related to the South American blue-black grassquit (*Volatinia jacarina*), that all 14 species plus the grassquit should be considered congeners within a new genus, *Geospiza*. This scheme is unlikely to be adopted pending a full evaluation of Steadman's arguments. However, there has been a more generally accepted change at the higher taxonomic levels of the 'finches and sparrows'. Lack kept his group as the subfamily Geospizinae, within the very large family Fringillidae. Today most taxonomists follow the 'Basel sequence', including the cardueline finches in the Fringillidae, while Darwin's finches end up as the 'fifth group' of the subfamily Emberizinae, within the bunting family Emberizidae. The sub-family Geospizinae is not generally recognized in this rearrangement (Paynter, 1970; Steadman, 1982). As the reader may have gathered, the taxonomy of the 'finches' is still in a state of flux (Paynter, 1970). We feel that as a matter of convenience and historical precedence, there is a good argument for retaining a subfamily or supergenus along the lines of the Geospizinae for the time being, and within that group, Lack's four genera and 14 species.

GALÁPAGOS GIANTS

Top left: Prickly-pear (*Opuntia*) tree (credit R. Leacock)
Top right: Tortoise (credit R. Leacock)
Middle: Land iguana soliciting cleaning (ectoparasite removal) from a *G. fuliginosa* perched on its back (credit F. Pölking)
Bottom: Marine iguana (credit R. Leacock)

p. 20. It is difficult to update completely Lack's distribution lists for Darwin's finches. Much confusion surrounds historical records, sight identifications are unreliable, and there is increasing evidence for interisland wandering and recent extinctions of some forms. Interested readers are referred to Harris (1973), Steadman (1981), and Sulloway (1982*a*, *b*) for details. *G. magnirostris* has since been found on Floreana (Charles) by Bowman (1963). However, Sulloway (1982*b*) concludes that while there is no doubt that Darwin's extinct 'giant *magnirostris*' was found on San Cristobal (Chatham) and Floreana, Bowman's (1963) specimen is not big enough to represent this subspecies; it almost certainly was an immigrant of the subspecies found throughout the central archipelago. Steadman (1981) found abundant fossil evidence on Floreana for *G. magnirostris* and the extinct Floreana mockingbird. *G. magnirostris* was also probably resident on Fernandina (Narborough) and Santa Fé (Barrington) at one time. *G. difficilis* has not been seen on Isabella (Albemarle), is extinct on Santa Cruz (Indefatigable), but is abundant on Fernandina (D. Schluter, personal communication). Steadman (1981) found fossils of *G. difficilis*, together with fossils of two apparently undescribed species of finch, in Cueva de Kubler on Santa Cruz. Sulloway (1982*b*) shows that *G. difficilis* also occurred on Floreana in Darwin's time; these early specimens were referred to as *G. nebulosa*, leading to Sulloway's claim that *G. difficilis* should be renamed *G. nebulosa* for reasons of nomenclatural precedence. Harris (1973) notes that *G. scandens* appears extinct on Pinzón (Duncan), and that there are recent records for *G. conirostris* on Wenman (in 1978 we saw some *G. magnirostris* but no *G. conirostris* there). *C. psittacula* has been seen recently on Santa Fé, but is thought to be extinct on Pinzón; *C. pallidus* has also been seen recently on Santa Fé and Pinta (Abingdon). *C. parvulus* is very rare on Pinta (D. Schluter, personal communication).

pp. 21–2. Lack's observations on stragglers have been supported by subsequent work. Dispersal of juveniles, often during the dry season of their first year, seems the most important source of interisland movements of finches. For instance, Daphne Major was considered by Lack, correctly, to have only two resident species, *G. fortis* and *scandens*. But since 1973, up to 50 *G. fuliginosa* or *magnirostris* have been seen there at once, and smaller numbers of *C. parvulus, psittacula, crassirostris*, and *Certhidea* have also been noted, as well as some *G. fortis* from neighbouring Santa Cruz (Grant *et al.*, 1975; Boag, 1981; Price and Millington, 1982). Few of these immigrants have remained on Daphne, or bred there. Thus, although interisland movement of finches may be regular, successful colonization or integration into existing populations may still be very rare. No new colonization has been documented since Darwin's visit in 1835.

pp. 22–3. Darwin did not appreciate that finches from different islands might belong to different but closely related species. He did not begin to separate his skins of geospizines by island after leaving Charles (Floreana), contrary to Lack's comments here. Sulloway (1982*a*) details how Darwin's initial misconception that the various finches belonged to three or four subfamilies led to the labelling problem and how Darwin attempted to assign localities to his skins after he returned to England, by using the private collections of shipmates aboard the *Beagle*. Steadman (1981) has begun to search for fossil remains of Galápagos vertebrates, particularly in the extensive lava tube systems of the islands of Santa Cruz and Floreana. The hunt for fossil or sub-fossil remains of Galápagos vertebrates has barely begun, but because predators such as the short-eared owl and barn owl often accumulate pellets and other remains in caves and lava tubes, such material may be more common than previously thought (Grant *et al.*, 1975).

CHAPTER III

p. 25. Lack's reference to the absence of a 'lark finch' on the open highlands may not be so curious. Most ecologists today are more hesitant to assume that evolution is always capable of filling 'vacant niches'. More importantly, there is some doubt as to how long the open highlands have existed (see Colinvaux (1972) for climatic fluctuations in the last few millenia); they may represent a man-made subclimax created by fire, domestic animals, and introduced plants (Wiggins and Porter, 1971).

pp. 25–8. Lack's assignment of species to habitats is generally correct, but even today systematic censuses, particularly along altitudinal gradients on large islands, are lacking (but see Sammalisto, 1966; Abbott *et al.*, 1977; Schluter, 1982*a*, *b*). Recent studies of the distribution of *G. difficilis* and *G. fuliginosa* on Pinta show that the two species are not confined to the humid and arid zones, respectively, but overlap in the transition zone (Schluter, 1982*a*) (see Notes for Chapter XV).

pp. 28–30. This section provides insight into Lack's future interests, culminating with the publication of *Ecological Isolation in Birds*, in 1971. Already Lack considered that 'differences in habitat in different regions are obviously correlated with the presence or absence of related species'. This conviction remained with him for the rest of his life (cf. Grant, 1977).

pp. 31–6. Much of this material was more fully developed in Lack's earlier monograph (Lack, 1945 L.). Taken together with Orr (1945 L.), these works remain an excellent and largely correct description of the general breeding biology of Darwin's finches.

pp. 32–3. Quantitative data now available (Downhower, 1978; Grant and Grant, 1980*a*; Boag, 1981; Grant, 1982) suggest that egg characteristics and clutch or brood sizes vary more than Lack implies, both inter- and intra-specifically. The latter part of this section is another portent of Lack's later research on the evolution of clutch size, summarized in the 1968 book, *Ecological Adaptations for Breeding in Birds*. Note particularly the remarkable sentence beginning 'But it is extremely difficult to see...'. This emphasis on individual selection and perhaps even the possibility of kin selection (later formalized by Hamilton (1964)), puts Lack ahead of his time. Contrast this with a comment on these pages made by Sir Julian Huxley in a letter dated 14 March 1946 offering Lack suggestions for the proofs of *Darwin's Finches*: 'Natural selection *in this case* operates as you say; but of course it *can* operate for the good of the species' (Lack archives, No. 51).

pp. 33–4. Lack's description of the breeding season is essentially correct. It was apparently based mostly on discussions with Mr Kubler, his host on Santa Cruz in 1939, as Lack's expedition journal contains many of the descriptions presented here. More detailed observations (Downhower, 1976, 1978; Grant and Grant, 1980*a*; Boag, 1981) show that breeding in the *Geospiza* is opportunistic and closely tied to the abundance and pattern of rainfall. Given the importance of rainfall to the ecology and evolution of the finches, there exists disappointingly little climatic information for the Galápagos even today (e.g. Alpert, 1963; Grant and Boag, 1980; Boag, 1981). What information is available suggests that rainfall can vary markedly, vertically with altitude on large islands, horizontally between islands, and temporally between months, years, or decades. The impact of this unpredictable climate on the ecology of the finches and other Galápagos organisms warrants much further study (Lack, 1950; Beatley, 1974).

Recent data also confirm that *G. scandens* and *conirostris* can breed before other *Geospiza*, and sometimes in years when other species do not breed at all (Grant and Grant, 1980*a*; Boag, 1981). This is because these species can exploit *Opuntia* cactus flowers, which bloom prior to the rainy season.

Snow (1966) has confirmed that moult in the finches can be intermittent, capable of being arrested if late rainfall triggers additional breeding.

pp. 34–6. Lack probably underestimated the importance of predation on the finches. Experiments have shown that finches recognize potential predators as such (Curio, 1969). Predation, particularly of nests and fledglings, can be heavy. Mockingbirds can be serious egg predators in some populations (Downhower, 1978; Grant and Grant, 1980*a*). On Pinta, nestlings are occasionally taken by hawks (D. Schluter, personal communication). Fledglings and adult finches are taken by mockingbirds, the short-eared owl, the Galápagos hawk, the common egret and the night and lava herons. Feral cats also take finches, and together with other introduced mammals, may be partly responsible for the decline of some native birds, such as the Hawaiian petrel, *G. difficilis*, and doves on Santa Cruz. The reference to Captain Colnett finding dead young (see also the reference to sterile eggs on p. 32) probably illustrates the opportunistic breeding behaviour of the finches. When a brief rainfall initiates breeding, but does not continue long enough to produce sufficient insect or fresh plant material to rear young, large numbers of eggs or young are likely to be abandoned. Modern workers have also noticed diseased feet in the finches, particularly on inhabited islands. This may represent an introduced avian pox disease. Small, imbedded cactus spines may also produce sores observed on the feet, or around the eyes and beak.

CHAPTER IV

pp. 36–40. Apart from a few comments by Bowman (1961), Curio and Kramer (1965) and Snow (1966), the topic of female plumage has remained unstudied. Lack was incorrect in suggesting that female plumage does not vary among populations of *Geospiza magnirostris*, *G. fortis* and *G. fuliginosa*. For example, immigrants of these species to Daphne Major island from neighbouring islands are more olive in colour than resident Daphne birds (Boag, 1981). Other populations of these species probably exhibit subtle local variation.

p. 44. Bowman (1961) concurred with Lack that female (and male) plumage is adaptive in reducing the probability of predator detection. The concealing coloration hypothesis is difficult to reconcile with the fact that in Darwin's finches, the nests and breeding displays are extremely conspicuous to would-be predators such as mockingbirds, hawks and owls (Downhower, 1978; Grant and Grant, 1980*a*).

CHAPTER V

pp. 44–5. Orr (1945 L.) described the generalized courtship and aggressive displays of captive ground finches (*Geospiza*) and confirmed that there were no interspecific differences in the postures used. He also described a 'sway display' apparently not seen by Lack. Field observations confirm Orr's descriptions (Ratcliffe, 1981). The wing-spread is a fundamental component of aggressive and sexual signalling. Wing-fluttering tends to be associated more with sexual signalling and food-begging than with aggression. Few passerine species exhibit such similarity in their sexual and aggressive displays (but see Baptista, 1978). The absence of interspecific differences in displays is also unusual (e.g. Crook, 1964; Zann, 1976). The displays of the tree finches (*Camarhynchus* spp.) and the warbler finch (*Certhidea*) have not been studied.

pp. 45–6. Lack actually spent more time collecting data on displays and vocalizations than is evident from this condensed section. He describes (onomatopoeically) the songs of most of the finch species, including island variants, as well as the whistle and call notes, in the earlier (1945 L.) California Academy paper. He was particularly struck with the difficulty of identifying individuals to species by ear. The high levels of intraspecific variation in song and of interspecific song overlap have been confirmed sonagraphically (Bowman, 1979, 1983). Intraspecific song variation is due mainly to song type differences present in all species, and to island dialects, which grade from being slight to pronounced. In the ground finches (*Geospiza*), song overlap between species occurs mainly between *G. fuliginosa* and *G. fortis*, although other species can sound very similar too (Ratcliffe, 1981). Bowman (1979, 1983) has suggested that the island dialects reflect adaptations to acoustic transmission characteristics on different islands, and that song overlap is due to vocal convergence in the same habitat. Using different analytical methods, Ratcliffe (1981) showed that dialect structure is also influenced by geographic isolation and genetic affinity. Song overlap in sympatric species may be due more to genetic affinity than to convergence. The significance of the song-type variation is still being studied (Grant and Grant, 1979, 1983).

Bowman (1979) suggested, and Ratcliffe (1981) confirmed that song functions in species recognition in the finches. Males are equally aggressive to playback of different song types, but respond more weakly to the playback of foreign dialect than local song. Males have no difficulty distinguishing between songs of their own species and similar-sounding songs of other species, and there is no interspecific territoriality. Whether females use song to recognize potential mates remains to be seen. Lack uses a 'loss-of-contrast' hypothesis to explain the low degree of species-specificity in geospizine song. This may be the correct explanation in some mainland–island comparisons within a single species (e.g. Lack and Southern, 1949; but see

PLATES

I. Above: *Camarhynchus psittacula*, the large insectivorous tree-finch (credit: J. Gould)
Below: *Camarhynchus crassirostris*, the vegetarian tree-finch (credit: J. Gould)

II. Above: Coastal scene (credit: R. Leacock)
Below: Nest of *Geospiza magnirostris* in *Opuntia* (credit: T. W. J. Taylor)

III. Above: Open uplands of Indefatigable (Santa Cruz) (credit: T. W. J. Taylor)
Below: Humid forest with trees of *Scalesia pedunculata* (credit: P. T. Boag)

IV. Above: *Geospiza fuliginosa*, the small ground-finch (credit: J. Gould)
Below: *Geospiza fortis*, the medium ground-finch (credit: J. Gould)

V. Above: *Geospiza magnirostris*, the large ground-finch. Female left, male right (credit: J. Gould)
Below: *Geospiza scandens*, the cactus ground-finch. Female top, male below (credit: J. Gould)

VI. (i) *Geospiza magnirostris* (credit: P. T. Boag); (ii) *Camarhynchus parvulus* (credit: P. T. Boag); (iii) *Camarhynchus pallidus* with its stick (credit: R. Perry); (iv) *Camarhynchus crassirostris* (credit: D. Lack); (v) the Indefatigable (Santa Cruz) mockingbird, *Nesomimus parvulus* (credit: R. Leacock); (vi) female *Geospiza fuliginosa* carrying nest material (credit: P. T. Boag)

VII. Beak variations in four species of *Geospiza* about 6/5 natural size; (i) *G. magnirostris*: left, on Tower (Genovesa) – right, on Jervis (Rábida); (ii) *G. fortis* on Charles (Floreana); (iii) *G. fuliginosa* on Chatham (San Cristóbal); (iv) *G. conirostris* on Hood (Española) (credit: California Academy of Sciences)

VIII. Above: *Certhidea olivacea*, the warbler-finch (credit: J. Gould)
Below: The Chatham (San Cristóbal) mockingbird, *Nesomimus melanotis* (credit: J. Gould)

PLATE I

Plates I, IV, V and VIII are available in colour as a download from www.cambridge.org/9780521272421

PLATE II

PLATE III

PLATE IV

PLATE V

PLATE VI

(i) (ii) (iii) (iv) (v) (vi)

PLATE VII

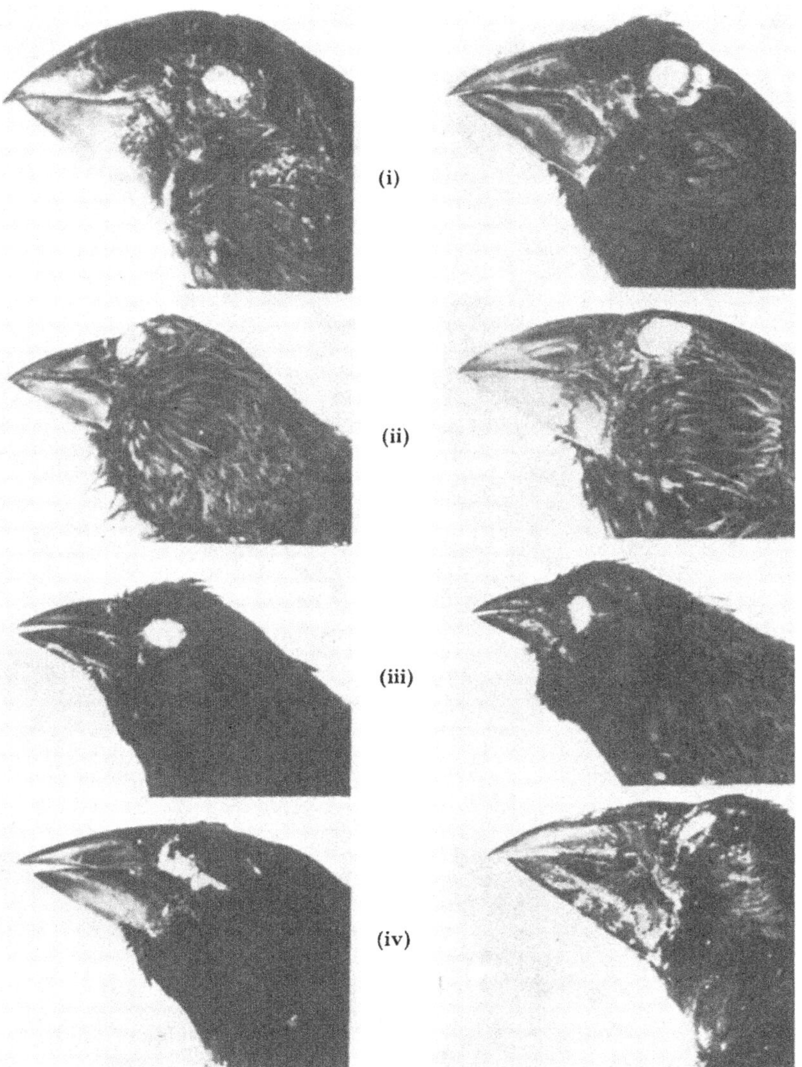

PLATE VIII

Thielcke, 1973; Becker, Thielcke and Wüstenberg, 1980). However, it is not clear why the songs of different species of sympatric Darwin's finches should lack distinctiveness. One possible explanation is that selection for vocal distinctiveness has been weak because of other well-developed recognition cues, such as visual signals (Ratcliffe, 1981).

pp. 48–9. Bowman (1961) found no correlation between the degree of skull pneumatization and blackness of plumage in the finches; even mature males occasionally have incompletely ossified skulls. There is a large amount of inter-individual variation in the age of acquisition of black plumage, but whether the proportion of males breeding in brown plumage actually differs among populations is not known. On Daphne Major, Genovesa and Pinta, where long-term breeding studies with marked birds have been conducted, few males breed in brown plumage (Boag, 1981; Grant and Grant, 1980*a*; D. Schluter, personal communication). The loss-of-contrast hypothesis, which Lack uses to explain the progressive loss of blackness, would predict that sexual selection on male plumage is weak or absent. Whether males in full black plumage have enhanced dominance or increased reproductive success compared to blackish or brown males is being studied on Daphne Major and Genovesa. Lack thought black males bred earlier, because they bred at higher altitudes on Santa Cruz. Bowman (1961) disagreed with this observation. Another hypothesis to explain the retention of juvenile or female plumage by breeding males is that brown males may acquire territories by deceiving neighbouring males about their status (e.g. Rowher, Fretwell and Niles, 1980). Whether black finches are less aggressive towards brown ones needs to be studied.

pp. 53–5. Ratcliffe (1981) repeated Lack's experiments on species recognition, using stuffed mounts; in all species of ground finches, males and females recognized conspecifics by visual cues. However, the beak by itself is not sufficient for species recognition; other stimuli from the head and body are also important. Visual recognition is also important in mate recognition, as Lack thought. Males from sympatric populations are better able to distinguish conspecific female models from similar-looking heterospecific ones than are allopatric males. Lack concluded that interspecific morphological differences among the finches reflected primarily adaptations for feeding specialization. The differences were only secondarily employed as species recognition cues. This seems to be the case, as there is no evidence of consistent intraspecific assortative mating by morphology in two populations studied to date (Grant and Grant, 1979; Boag, 1983).

CHAPTER VI

pp. 55–60. This chapter represents a radical revision of Lack's 1945 monograph, in which he explained many morphological differences between the finches as nonadaptive. The rather sketchy description of the functional relationships between morphology and foods eaten has been confirmed and extended by the exhaustive descriptions of Bowman (1961), and quantitative analyses of *Geospiza* diets by Abbott *et al.* (1975, 1977), Grant *et al.* (1976), Smith *et al.* (1978), Grant and Grant (1981), Grant (1981*b*), Boag and Grant (1981), and Schluter (1982*a*, *b*). DeBenedictus (1966) describes one way Darwin's finches get at buried seeds, while Bowman and Billeb (1965), Carpenter (1966), MacFarland and Reeder (1974), and Boag (1981) catalogue such interesting non-seed feeding behaviours as blood-drinking and consumption of reptile ectoparasites and autotomous lizard tails. There remains a pressing need for similar quantitative work on the other finch genera (cf. Smith and Sweatman, 1976).

pp. 60–4. This section is another harbinger of Lack's later emphasis on the role of competition in structuring avian communities (Lack, 1971). As discussed in our Introduction, he changed his ideas during the early years of World War II; his views were certainly well developed in time for the Easter 1944 meeting of the British Ecological Society when he, Varley, and Elton argued in favour of the importance of competition (Anon., 1944; Lack, 1944 L.; Elton, 1946). The idea took some time to catch on; it was not readily accepted at that meeting, and much later Bowman (1961) still adhered to the views of Andrewartha and Birch (1954), arguing that competition was not evident in his studies of Darwin's finches. After several years of general acceptance, a debate about competition has surfaced again (Wiens, 1977; Abbott *et al.*, 1977; Diamond, 1979; Strong, Szyska and Simberloff, 1979; Grant and Abbott, 1980; Simberloff and Connor, 1981; Simberloff and Boecklen, 1981). These issues are far from being resolved, but it has become obvious that simple inferential descriptions of the way competition works in structuring avian communities, the approach used by Lack, will gradually have to make way for more rigorous tests of the theory (Grant, 1977).

pp. 64–6. Some of the examples given in this section have since been more fully described by Lack (1971). See also Grant (1972*a*, 1978) for an interesting addendum on Norfolk Island *Zosterops*. Note that in this section Lack refers to Huxley's (1942 L.) use of Gause's (1934 L.) theory in discussing food differences between tern species. Given Lack's familiarity with Huxley and his work, this was presumably an important specific source of inspiration in the early 1940s (Jackson, 1981).

pp. 66–70. Grant and Grant (1982) have re-examined the *conirostris–magnirostris* comparison on Genovesa (Tower) and Española (Hood). They conclude that many of Lack's ideas are supported in this instance, although it remains unclear why *G. fortis* is absent from Genovesa and *magnirostris* from Española. D. Schluter has also examined *difficilis* on several of the islands mentioned in this section, again largely confirming Lack's interpretations, but with many more data on diets and the availability of foods (Schluter, 1982*a*; Schluter and Grant, 1982).

pp. 70–2. Virtually no detailed quantitative ecological studies have been carried out on *Camarhynchus* or *Certhidea*. The little information available (e.g. see Bowman, 1961, 1963) does not point to any obvious problems with Lack's discussion of this group, but does underline the strong need for further work on the tree and warbler finches.

CHAPTER VII

pp. 72–4. Lack's interest in examining patterns of morphological variation marks a significant step forward in the quantification of evolutionary ecology (cf. Lack, 1965). His papers contain detailed notes on Miller's (1941 L.) work on *Junco*, and despite the paucity of studies in that area he seemed to consider it very important. In a letter dated 23 April 1940 (Lack archives, No. 48), R. A. Fisher responded to Lack's requests for statistical advice by explaining the distinctions between standard deviations, standard errors, and sums of squares, and how to decide how many significant digits to cite in his tables of data. Most of the measurement data were collected in 1939 at the California Academy of Sciences in San Francisco.

pp. 74–9. This section illustrates Lack's indecision about the adaptive significance of subspecific morphological variation. In the preface to the 1961 reprint of *Darwin's Finches* (Lack, 1961), he indicated that he had come to view these too as adaptive, and thus one can follow a progression from a general acceptance of the significance of generic differences in 1945, to species differences in 1947, and subspecific differences in 1961. This trend is partly explained by Lack's gradual acceptance of significant differences between islands in aspects of their environment relevant to finch ecology, such as the quantity and quality of food. This concept, originally championed by

Bowman (1961), and partially confirmed (as the 'floristics hypothesis') by Abbott *et al.* (1977), was eventually recognized by Lack (1969) himself.

CHAPTER VIII

pp. 81–6. This section, particularly Fig. 17, has been one of the most frequently cited parts of Lack's book, in works ranging from introductory biology textbooks to specialized theoretical treatises. The idea that inter-specific competition could mould the phenotypic distributions of a series of closely related species has proved an intuitively appealing concept. The ideas here played a major role in stimulating the plethora of work destined to appear 20 years later on topics such as limiting similarity, character displacement, and Hutchinsonian morphological ratios, among others (Hutchinson, 1957; MacArthur, 1972; Grant, 1972*b*; Horn and May, 1977; but see Simberloff and Boecklen, 1981).

p. 83. The extreme variability of *G. fortis* on Santa Cruz (Indefatigable) has been examined by Snow (1966), who attributed it to hybridization, and by Ford, Parkin and Ewing (1973), who concluded that the almost complete gradation between *fortis* and *magnirostris* at Academy Bay may represent either hybridization or disruptive selection. *G. fortis* is certainly the most variable finch species, but the Academy Bay population is unique in this respect, and may demand an explanation that does not necessarily apply to other populations of the same species.

p. 85. In 1945, Lack considered the Daphne and Hermanos (Crossman) birds to be hybrids between *fortis* and *fuliginosa*, and then changed his opinion here. Much later (1969) his uncertainty about them surfaced again, although he still saw them primarily as examples of what we would now refer to as character release (Grant, 1972*b*). Harris (1973) thought most Daphne individuals were assignable to either *fuliginosa* or *fortis*. Recent field work has shown that *fortis* on Daphne are still of intermediate size (contrary to a statement by M. Williamson (1981)), that *fortis* and *fuliginosa* do occasionally immigrate to Daphne from neighbouring Santa Cruz, and that Daphne *fortis* sometimes mate with immigrant *fuliginosa*, producing offspring of intermediate size (Grant *et al.*, 1975; Boag, 1981). This introgression seems to result in part from reduced selectivity by female *fuliginosa* unable to find males of their own species, and in part from poor species recognition in Daphne *fortis*. *Geospiza* use morphology as a species recognition cue and small Daphne *fortis* resemble immigrant *fuliginosa* (Ratcliffe, 1981). But the situation on Daphne is by no means a simple case of introgression between two species; the known selection regime in one year, 1977, favoured larger, not smaller birds (Boag and Grant, 1981), and there remains evidence for character release in the form of diet expansion by *fortis* in periods of food shortage (Boag, 1981). The more this population is studied, the less it seems appropriate to describe it as being the product of a single evolutionary process, and hence its continued widespread citation as a clear-cut example of competition-mediated character release is probably not correct.

pp. 86–7. This section is one of the first to apply simple multivariate techniques to the analysis of avian morphology. However, Lack's results are questionable, since more recent analyses (Boag, 1981) show that very strong positive allometry does exist for characters such as bill depth in *fortis*, while the same character in species such as *fuliginosa* or *magnirostris* appears to vary isometrically with overall body size. Huxley's book (1932; incorrectly cited by Lack as 1927 L.) remains a classic in this rather understudied field; more recent applications of allometric techniques are found in Gould (1966) and Lande (1979).

p. 90. Here Lack points to the need to study *Geospiza* diets in more detail, particularly in the dry season, a comment that foreshadows the conclusions of recent work such as that by Smith *et al.* (1978) and Grant and Grant (1980*b*). The dry season, especially before plants such as the *Opuntia* cactus bloom, is the most critical time of the year ecologically for the *Geospiza*. It is surprising that Lack did not elaborate further on these points; he did, for instance, in his raptor paper (1946). When he wrote the latter, *Darwin's Finches* was largely completed, as he cites the book as being published in 1946. There he suggested that part of the reason he took so long to accept species differences in finch feeding habits was the fact that most of his time had been spent in the Galápagos during the wet season, when most finch species had similar diets, given a superabundance of insects and fresh vegetation. Bowman (1961) apparently missed this point: he did most of his field work in the wet season as well, and he remarked that species in the same environment were often seen eating the same foods. This became one of his arguments against the existence of competition in the finches.

CHAPTER IX

pp. 91–2. In a letter dated 23 April 1940 (Lack archives, No. 48), R. A. Fisher advised Lack to examine the relationship between population size and morphological variability by estimating relative population sizes using island areas, and then regressing a measure of morphological variability on those population estimates. In the same letter he pointed out that Darwin had anticipated Sewall Wright in suggesting that small, isolated populations tend to be less variable than larger ones. Lack's failure to find an obvious relationship between population size and variation is not surprising, as Fisher's theoretical expectation holds primarily for an idealized population. The samples of birds found in museums usually consist of several lots of birds collected at different places and times, by different workers (Grant and Price, 1981). The heterogeneity among such samples is often sufficient to hinder accurate estimates of true population variation.

pp. 93–4. The relative variability of an entire species is less of a problem, and Lack's data have stimulated further studies in this area (Ford *et al.*, 1973; Grant *et al.*, 1976; Abbott *et al.*, 1977). *G. fortis* has particularly large coefficients of variation for its bill dimensions. This seems partly related to its generalist diet and probably also to occasional introgression with *G. fuliginosa* and *G. magnirostris* on islands such as Daphne and Santa Cruz, respectively. Further work is needed to confirm these suspicions (e.g. Yang and Patton, 1981). Van Valen (1965) reconsidered the possibility of island populations showing increased variability due to competitive release, stimulating an ongoing debate on the 'niche variation' hypothesis (Rothstein, 1973; Grant and Price, 1981).

CHAPTER X

There continue to be few reports of either interspecific or intergeneric hybridization, even in well-studied populations of Darwin's finches (reviewed by Bowman, 1983). On Genovesa, P. R. and B. R. Grant found two *Geospiza conirostris* × *G. magnirostris* pairs between 1978 and 1981. Boag (1981) has documented successful hybridization between *G. fortis* and *G. fuliginosa* on Daphne Major and this hybridization continues at a low annual rate (2–4% of breeding *fortis* pairs involve *fortis–fuliginosa* matings, usually with a *fuliginosa* female). There has also been a successful pairing between a *G. scandens* and *G. fortis* on Daphne (T. Price, personal communication). The breakdown of ethological isolation can be attributed, in some cases, to accidental fostering of young by heterospecific foster parents during the nestling or fledgling stage (Bowman, 1983). Occasional heterospecific egg 'dumping' also occurs (D. Schluter, personal communication) but its effects on mate selection are unknown. The Daphne Major situation is somewhat more complex, since the hybridizing females are frequently immigrant

G. fuliginosa from Santa Cruz. In this case, hybridization may be due to a lowering of the mating threshold in the absence of conspecific mates, such as that suggested for *Vermivora* warblers (Gill, 1980).

pp. 96–7. The populations with intermediate morphology do, in some cases, have intermediate feeding niches, as Lack suggests. *G. conirostris* has a broad feeding niche on Española, taking foods eaten by *G. fortis*, *G. scandens*, and *G. magnirostris* on other islands (Grant and Grant, 1982). *G. darwini* on Culpepper (Darwin) is currently considered a form of *G. magnirostris* with an unusually variable bill.

CHAPTER XI

pp. 100–6. The precise systematic relationships of the finches remain largely a matter for skilled speculation even today. One problem has been the lack of an identifiable ancestor (but see below). Another problem has been the lack of an independent set of characters for classification, separate from those important in feeding ecology. Preliminary examination of syringeal musculature (Cutler, 1970) and karyotypes (Jo, 1976) seem to confirm the overall integrity of the group and its affinities with the nine-primaried oscines. An initial biochemical study (Yang and Patton, 1981) produced phylogenetic trees very much like those suggested by Lack here. These electrophoretic data confirm the clusters of species within the *Geospiza* and *Camarhynchus* genera, and suggest that *Certhidea* is sufficiently distinct from both, that it might be the product of a separate invasion or a very early divergence from the ancestral stock. Ratcliffe (1981) has shown that there exists a correspondence between matrices of genetic, morphological, and song interpopulation similarities, which is further evidence supporting Lack's phylogenetic tree.

In the search for the ancestor to Darwin's finches, Steadman (1982) suggests that the Cocos finch was independently derived from the northern subspecies of grassquit, *Volatinia jacarina splendens*, while the finches in the Galápagos are derived from *V. j. peruviensis*. The main reason for this suggestion is that the Cocos finch and *splendens* share plumage characters such as an irregular, whole body moult into mature, black male plumage, while most Galápagos finches and *peruviensis* moult into black from head to tail. Less convincingly, Bowman (1983) still identifies *Melanospiza richardsoni* as a modern relative of the ancestor. F. Sulloway and E. Mayr have indicated to us that they feel Steadman's suggestion is likely to prove the correct one (personal communication). The overall taxonomy of the 'finch' group of passerines is sufficiently controversial to warn against the acceptance of specific taxonomic proposals too hastily. But Steadman's efforts represent the first serious attempt to solve the fascinating puzzle of where Darwin's finches did come from, and are bound to stimulate further work on understudied candidates such as *Volatinia* (Carvalho, 1957; Alderton, 1963; Murray, 1982). Raikow (1976) recently provided a very thorough study of the same problem in the Hawaiian sicklebills (Drepanididae), using myological characters to produce a convincing picture of their ancestry.

CHAPTER XII

pp. 107–10. The land-bridge hypothesis has since been rejected in favour of an independent origin of the islands. Potassium–argon dating gives the islands an upper age limit of about five million years (Bailey, 1976), although layering of lava flows and sea-level fluctuations (Connor and Simberloff, 1978) make it difficult to estimate biologically meaningful ages for particular sites. Other references on the geology of the islands can be found in Paepe (1955), Williams (1966) and McBirney and Aoki (1966). Volcanic activity continues in the western part of the archipelago, on Fernandina and Isabela, the last eruption being in 1979.

pp. 111–12. Since Lack's visit, only one more species of land bird, the paint-billed crake (*Neocrex erythrops*) has become established in Galápagos (Harris, 1974). Very little work has been done comparing island races, subspecies or species of birds with mainland relatives (see Lanyon, 1978 for an exception). For a discussion of the possible mainland ancestors of Darwin's finches, refer to the Notes for Chapter XI.

CHAPTER XIII

This is an important chapter in which Lack attempts to sort out adaptive versus non-adaptive variations in the finches. One of his main difficulties in this regard was his failure to appreciate inter-island differences in climate, physiognomy and flora. He was therefore left trying to explain inter-island population variation as the result of non-adaptive phenomena, such as genetic drift and founder effect.

p. 117. Some information is now available on the inheritance of several phenotypic characters in the finches (Boag and Grant, 1978; Grant *et al.*, 1979; Yang and Patton, 1981; Grant, 1981*c*; Boag, 1983).

pp. 119–21. Peripheral populations of Darwin's finches not only show the most extreme morphological divergence (Hamilton and Rubinoff, 1967; Power, 1975), they also tend to have the most distinct vocal dialects (Ratcliffe, 1981). Lack's Fig. 22 has become particularly well known. Note that recent data change the percentages for Chatham and Charles to 31 % and 30 % respectively (Sulloway, 1982*a*).

p. 121. Populations of Darwin's finches which are separated by water gaps only a few kilometres wide may be quite isolated. The only records of inter-island movement by marked individuals are those of three *G. scandens*, which were banded as juveniles on Daphne Major and found a short time later on Daphne Minor, approximately six kilometres to the north (Grant *et al.*, 1980), and two *fortis* banded on Daphne Minor as juveniles, which later turned up on Daphne Major (T. Price, personal communication).

pp. 122–4. Here Lack discusses the likelihood of genetic drift and founder effects being important in the allopatric divergence of populations. There is some evidence (Boag, 1983) that effective *Geospiza* population sizes on islands such as Daphne Major may sometimes fall as low as 100 individuals. However, in most cases, even weak gene flow or selection is likely to mask such sampling effects. Sewall Wright (1980) has pointed out that Huxley (1942 L.) and others (including Lack here) have misconstrued his ideas on the evolution of small populations. Wright viewed the 'Sewall Wright effect' primarily as a process by which small population sizes facilitated shifts between 'adaptive peaks'; his theory emphasized selection and pleiotropy, as well as sampling phenomena such as genetic drift. The non-adaptive explanation for allopatric differentiation which Lack eventually favoured, here ascribed to Muller (1940 L.), is no longer considered viable.

CHAPTER XIV

pp. 125–30. The model of allopatric or geographic speciation described here has been widely accepted by evolutionary biologists as an appropriate one for most vertebrates (Mayr, 1966; Bush, 1975; White, 1978). Other models of speciation (e.g. parapatric, sympatric) may be more appropriate in certain continental situations (White, 1978; Templeton, 1981). There is a great deal of controversy about the importance of natural selection against hybridization in reinforcing reproductive isolating mechanisms, mainly because there are few well-documented examples of reproductive character displacement (Paterson, 1978; Waage, 1979; Templeton, 1981). The experiments on species recognition by Ratcliffe (1981) support the reinforcement hypothesis in Darwin's finches.

pp. 131–3. In discussing the kinds of variation present in natural populations Lack omits polymorphisms, which were little studied until the 1950s. Grant *et al.* (1979) found a nestling bill colour polymorphism in Darwin's finches, which may function in parent–offspring signalling. Many avian polymorphisms occur in plumage characters and could give rise, theoretically, to incipient sympatric speciation through positive assortative mating (O'Donald, Wedd and Davis, 1974; Cooke, 1978; Nuechterlein, 1981). Grant and Grant (1979) described a song type polymorphism in Isla Genovesa *conirostris* which, they suggested, could function to subdivide the breeding population. Confirmatory evidence of this is lacking. It is generally accepted that the adaptive radiation of the finches proceeded mostly, if not entirely, through the allopatric process (Grant, 1981*a*).

CHAPTER XV

pp. 134–6. Here is the essence of Lack's thinking on ecological isolation, which he developed in later papers (1945, 1946) and in the book *Ecological Isolation in Birds*. Readers will recognize the descriptions of fundamental and realized niches, which are basic concepts in modern ecology (Hutchinson, 1957). Lack talks about 'seasonal factors' affecting ecological isolation, but mainly with regard to migratory species. Later (1946) he stressed that during seasonal flushes in food abundance, such as those which occur in the Galápagos rainy season, the diets of sympatric species could be expected to overlap considerably. Competition for food would be manifested mainly in the food-limited dry season in such a case. Competition theory would predict that finch diets should diverge interspecifically under conditions of low food availability. These predictions were supported by a seasonal study of feeding behaviour by Smith *et al.* (1978).

pp. 136–7. In the last paragraph of this section, Lack mentions G. C. Varley, who emphasized the importance of knowing the patterns and sources of mortality in populations, and who made important contributions to the use of life tables and 'key factor analysis' in ecology. Professor Varley recently elaborated on the discussion which Lack cites here. His view was that density-dependent predation reduced the level of interspecific competition among populations at or near evolutionary equilibrium. When a population invaded a relatively predator-free environment (such as the Galápagos), this resulted in population expansion until food became a limiting factor. Increased intra- and inter-specific competition for food would then favour speciation through ecological isolation, through shifts in morphology or habitat preference. Thus Varley felt that the switch from population regulation by predators to population regulation by food was the crucial stimulus to speciation (cf. Lack's comments on p. 114 and his reference to Worthington (1940 L.)).

Lack goes on to cite examples of geographic, habitat, food and morphological separation of finch species. Some of these have since been examined in more detail. Relations among the tree finches (*Camarhynchus* spp.) however are still poorly known. The ecology of the mangrove finch (*C. heliobates*) has not been studied at all (see Curio and Kramer (1964) for some general observations). Where *G. conirostris* replaces *scandens*, it indeed subsumes all or part of the *scandens* niche (Grant and Grant, 1982), supporting the interspecific competition hypothesis. Schluter and Grant (1982) have concluded that the geographic replacement of *G. fuliginosa* by *G. difficilis* in the arid zones of Tower (Genovesa), Culpepper (Darwin) and Wenman (Wolf) is partly due to competitive exclusion, rather than to a lack of appropriate food resources for *fuliginosa* as Bowman (1961) suggested. On Pinta, *G. difficilis* and *G. fuliginosa* are not completely separated by habitat, being extensively syntopic year round (Schluter, 1982*a*). Interspecific competition may have had a significant influence on the two populations in the past, but at present the two do not compete for food to any measurable degree. The distribution of the two species along an altitudinal gradient on Pinta is best explained by the distribution of food resources. Finally, the hypothesis that different species of ground finches take different foods is well-supported by the studies of Bowman (1961), Grant *et al.* (1976), Abbott *et al.* (1977), Grant (1981*b*) and Schluter (1982*a*, *b*).

pp. 137–46. Whether interspecific competition has been as dominant a force in the structuring of animal communities as Lack suggests here is currently under debate (Connor and Simberloff, 1978, 1979; Grant and Abbott, 1980). Lack marshalled distributional data from a number of species to support the hypothesis. Formulating and testing alternative explanations for these distributions is quite difficult. In the case of Darwin's finches at least, a substantial data base still lends support to the competition hypothesis (Alatalo, 1982; Grant and Schluter, 1983).

CHAPTER XVI

p. 147. Lack's disinclination to accept a dichotomy between species formation and evolutionary adaptation is interesting in the light of the current debate over 'punctuated equilibria' (P. G. Williamson, 1981). Williamson, together with colleagues such as S. J. Gould, argue that fossil lineages for many taxa show little change over long periods of time, with occasional bursts of speciation. They claim that the gradualistic model of speciation suggested by Lack here, is inadequate to explain these observations. Many staunch supporters of Lack's view remain, however, and the issue is far from being resolved (Charlesworth *et al.*, 1982).

pp. 147–8. For recent work on the Canary Islands problem, see Grant (1979*a*, *b*). The cyclical pattern of speciation Lack advocates here (cf. Mayr, 1942 L.; Grant, 1981*a*) has become the classical explanation for insular adaptive radiation (an interesting modification of this is the 'taxon cycle' concept, e.g. Ricklefs and Cox, 1972). The main point of controversy about the process in the Galápagos, stimulated by Bowman (1961), is the relative importance of ecologically driven allopatric differentiation and competition mediated differentiation in sympatry. It is now accepted that ecological differences between islands are real (Lack, 1969; Abbott *et al.*, 1977), but there remains much controversy about what takes place in sympatry. Are communities built up largely by random processes, or does competition bias colonization rates, or alter phenotypes in sympatry by natural selection for character shifts (Grant, 1972*b*, 1981*a*; Grant and Abbott, 1980; Simberloff and Connor, 1981)?

p. 152. See Raikow (1976) for updated information on the sicklebills.

pp. 154–7. Perhaps one of the best modern examples of insular radiation in a non-avian group is Hampton Carson and his colleagues' intricate study of Hawaiian fruit flies (Drosophilidae). Up to 800 species may have rapidly differentiated from one or a few ancestral species (Carson and Kaneshiro, 1976; Carson, 1978), and the extensive knowledge of the genetics of these insects has permitted reconstruction of many sequences of speciation: see Gorman (1979) and M. Williamson (1981) for other examples of modern island studies. There remains much work to be done on other Galápagos organisms. Even such famous groups as the Galápagos tortoises are only now being subjected to the modern techniques of ecological genetics (Marlow and Patton, 1981). It is ironic that one of the most thoroughly studied Galápagos vertebrates is the introduced rat (*Rattus rattus*) (Patton, Yang and Myers, 1975; D. A. Clark, 1980; 1982, D. B. Clark, 1980). There are many excellent examples of radiations in plants (Wiggins and Porter, 1971), but much basic botanical work remains to be done in the islands, simply to complete distributional lists. Connor and Simberloff (1978) point out that the number of collecting expeditions visiting a given island remains one of the best predictors of the plant species diversity on that island.

p. 158. Lack's emphasis that interactions between species are an essential component of island radiations remained with him throughout the rest of his life (Lack, 1976; Grant, 1977). Later, this area of ecology was revolutionized by MacArthur and Wilson's (1967) theory of island biogeography, which Lack never really accepted (Grant, 1977).

pp. 158–9. Lack's concern over island faunas is as important today as it ever was. Many more people live on and visit the Galápagos now than when Lack was there. The pressures exerted on the indigenous Galápagos flora and wildlife are consequently large and continue to grow, despite the conservation efforts of the Ecuadorian National Park Service, with the assistance of the Charles Darwin Foundation and Research Station. Currently the greatest threat is from introduced rats and feral dogs, cats and livestock, which have devastated several reptile populations (e.g. Laurie, 1982) and at least one bird population (Coulter, Duffy and Harcourt, 1982). Many other vertebrate populations remain vulnerable (Harris, 1974). Less obvious but just as serious are the threats from habitat modification through agriculture and rural development, and the importation of plant and animal species from the mainland. Tourism has increased public awareness of the islands' value, and turned them into a net source of income for the Ecuadorian government, but it has also brought its share of sociological problems (Moore, 1980), and increased expectations for development. The reconciliation of the frequently disparate aims of development, scientific research, conservation and tourism continues to be a difficult task. Despite these problems, there have been several conservation successes, notably in the control of goats on several islands, and in the captive breeding and re-introduction of the Galápagos tortoise. Local public education in conservation is also being implemented. But at present, there is no room for complacency with regard to the future of Galápagos. Only the very strongest and continuous commitment to their preservation, by both the Ecuadorian government and the world conservation community will ensure the survival of this unique ecosystem.

MODERN REFERENCES

Abbott, I., Abbott, L. K. and Grant, P. R. (1975). Seed selection and handling ability of Darwin's Finches. *Condor*, **77**, 332–5.

Abbott, I., Abbott, L. K. and Grant, P. R. (1977). Comparative ecology of Galápagos Ground Finches (*Geospiza*, Gould): Evaluation of the importance of floristic diversity and interspecific competition. *Ecol. Monogr.* **47**, 151–84.

Alatalo, R. V. (1982). Bird species distributions in the Galápagos and other archipelagos: competition or chance? *Ecology*, **63**, 881–7.

Alderton, D. C. (1963). The breeding behavior of the Blue-black Grassquit. *Condor*, **65**, 154–62.

Alpert, L. (1963). The climate of the Galápagos Islands. *Occ. Pap. Calif. Acad. Sci.* **44**, 21–44.

Andrewartha, H. G. and Birch, L. C. (1954). *The Distribution and Abundance of Animals*. University of Chicago Press, Chicago.

Anon. (1944). British Ecological Society Easter Meeting 1944. Symposium on 'The Ecology of Closely Allied Species'. *J. Anim. Ecol.* **13**, 176–7.

Bailey, K. (1976). Potassium–argon ages from the Galápagos Islands. *Science*, **192**, 465–7.

Baptista, L. F. (1978). Territorial, courtship, and duet songs of the Cuban Grassquit (*Tiaris canora*). *J. Orn.* **119**, 91–101.

Beatley, J. C. (1974). Phenological events and their environmental triggers in Mojave desert ecosystems. *Ecology*, **55**, 856–63.

Becker, P. H., Thielcke, G. and Wüstenberg, K. (1980). Versuche zum angenommenen Kontrastverlust im Gesang der Blaumeise (*Parus caeruleus*) auf Teneriffa. *J. Orn.* **121**, 81–95.

Boag, P. T. (1981). Morphological variation in the Darwin's finches (Geospizinae) of Daphne Major Island, Galápagos. Unpublished Ph.D. thesis, McGill University, Montreal.

Boag, P. T. (1983). The heritability of external morphology in Darwin's Ground Finches (*Geospiza*) on Isla Daphne Major, Galápagos. *Evolution*, **37**, in press.

Boag, P. T. and Grant, P. R. (1978). Heritability of external morphology in Darwin's finches. *Nature*, **274**, 793–4.

Boag, P. T. and Grant, P. R. (1981). Intense natural selection in a population of Darwin's finches (Geospizinae) in the Galápagos. *Science*, **214**, 82–5.

Bowman, R. I. (1961). *Morphological differentiation and adaptation in the Galápagos finches*. University of California. *Publs. Zool.* **58**, 1–302.

Bowman, R. I. (1963). Evolutionary patterns in Darwin's Finches. *Occ. Pap Calif. Acad. Sci.* **44**, 107–40.

Bowman, R. I. (1979). Adaptive morphology of song dialects in Darwin's finches. *J. Orn.* **120**, 353–89.

Bowman, R. I. (1983). The evolution of song in Darwin's Finches. In

Patterns of Evolution in Galápagos Organisms, ed. A. E. Levinton and R. I. Bowman. Amer. Assoc. Adv. Sci. Special Publ. 1, Pacific Division, San Francisco.

Bowman, R. I. and Billeb, S. L. (1965). Blood-eating in a Galápagos finch. *Living Bird*, **4**, 29–44.

Bush, G. L. (1975). Modes of animal speciation. *Ann. Rev. Ecol. Syst.* **6**, 339–64.

Carpenter, C. C. (1966). The marine iguana of the Galápagos islands, its behavior and ecology. *Proc. Calif. Acad. Sci., Ser.* 4, **34**, 329–75.

Carson, H. L. (1978). Speciation and sexual selection in Hawaiian *Drosophila*. In *Ecological Genetics: the Interface*, ed. P. F. Brussard, pp. 98–101. Springer-Verlag, New York.

Carson, H. L. and Kaneshiro, K. Y. (1976). *Drosophila* of Hawaii: systematics and ecological genetics. *Ann. Rev. Ecol. Syst.* **7**, 311–45.

Carvalho, C. T. de (1957). Notas ecológicas sôbre *Volatinia jacarina* (Passeres, Fringillidae). Boletim do Museu Paraense Emilio Goeldi (Brasil). *Zoologia*, No. 2, pp. 1–10.

Charlesworth, B., Lande, R. and Slatkin, M. (1982). A Neo-Darwinian commentary on macroevolution. *Evolution*, **36**, 474–98.

Clark, D. A. (1980). Age- and sex-dependent foraging strategies of a small mammalian omnivore. *J. Anim. Ecol.* **49**, 549–63.

Clark, D. A. (1982). Foraging behavior of a foraging omnivore (*Rattus rattus*): meal structure, sampling and diet breadth. *Ecology*, **63**, 763–72.

Clark, D. B. (1980). Population ecology of *Rattus rattus* across a desert–montane forest gradient in the Galápagos Islands. *Ecology*, **61**, 1422–33.

Cody, M. L. and Diamond, J. M. (Eds.). (1975). *Ecology and Evolution of Communities*. Belknap Press, Cambridge, Mass.

Colinvaux, P. (1972). Climate and the Galápagos Islands. *Nature*, **240**, 17–20.

Connor, E. F. and Simberloff, D. (1978). Species number and compositional similarity of the Galápagos flora and avifauna. *Ecol. Monogr.* **48**, 219–48.

Connor, E. F. and Simberloff, D. (1979). The assembly of species communities: chance or competition? *Ecology*, **60**, 1132–40.

Cooke, F. (1978). Early learning and its effect on population structure. Studies of a wild population of Snow Geese. *Zeits. fur Tierpsychol.* **46**, 344–58.

Coulter, M. C., Duffy, D. C. and Harcourt, S. (1982). Status of the Dark-rumped Petrel on Isla Santa Cruz. *Notic. de Galáp.* **35**, 24.

Crook, J. H. (1964). The evolution of social organization and visual communication in the Weaver Birds (Ploceinae). *Behav. Suppl. X.* E. J. Brill, Leiden.

Curio, E. (1969). Funktionsweise und Stammesgeschichte des Flugfeinderkennens einiger Darwinfinken (Geospizinae). *Zeits. fur Tierpsychol.* **26**, 394–487.

Curio, E. and Kramer, P. (1964). Vom Mangrovefinken (*Cactospiza heliobates* Snodgrass and Heller). *Zeits. fur Tierpsychol.* **21**, 223–34.

Curio, E. and Kramer, P. (1965). On plumage variation in male Darwin's Finches. *Bird-Banding*, **36**, 27–44.

Cutler, B. D. (1970). Anatomical studies on the syrinx of Darwin's Finches. Unpublished Master's thesis, San Francisco State University, San Francisco.

DeBenedictus, P. A. (1966). The bill-brace feeding behavior of the Galápagos finch *Geospiza conirostris*. *Condor*, **68**, 206–8.

Diamond, J. M. (1979). Niche shifts and the rediscovery of interspecific competition. *Amer. Sci.* **66**, 322–31.

Downhower, J. F. (1976). Darwin's Finches and the evolution of sexual dimorphism in body size. *Nature*, **263**, 558–63.

Downhower, J. F. (1978). Observations on the nesting of the Small Ground Finch *Geospiza fuliginosa* and the Large Cactus Ground Finch *G. conirostris* on Española, Galápagos. *Ibis*, **120**, 340–6.

Duffy, D. C. (1981*a*). Ferals that failed. *Notic. de Galáp.* **33**, 21–2.

Duffy, D. C. (1981*b*). A master plan for ornithology in the Galápagos Islands. *Notic. de Galáp.* **34**, 10–16.

Eibl-Eibesfeldt, I. (1960). *Galápagos*. MacGibbon & Kee, London.

Elton, C. (1946). Competition and the structure of ecological communities. *J. Anim. Ecol.* **15**, 54–68.

Ford, H. A., Parkin, D. T. and Ewing, A. W. (1973). Divergence and evolution in Darwin's Finches. *Biol. J. Linn. Soc.* **5**, 289–95.

Gill, F. B. (1980). Historical aspects of hybridization between Blue-winged and Golden-winged Warblers. *Auk*, **97**, 1–18.

Gorman, M. L. (1979). *Island Ecology*. Halsted Press, London.

Gould, S. J. (1966). Allometry and size in ontogeny and phylogeny. *Biol. Rev.* **41**, 587–640.

Grant, B. R. and Grant, P. R. (1979). Darwin's finches: population variation and sympatric speciation. *Proc. natl. Acad. Sci.* **76**, 2359–63.

Grant, B. R. and Grant, P. R. (1981). Exploitation of *Opuntia* cactus by birds on the Galápagos. *Oecologia*, **49**, 179–87.

Grant, B. R. and Grant, P. R. (1982). Niche shifts and competition in Darwin's Finches: *Geospiza conirostris* and congeners. *Evolution*, **36**, 637–57.

Grant, B. R. and Grant, P. R. (1983). Fission and fusion in a population of Darwin's Finches: an example of the value of studying individuals in ecology. *Oikos*, **34**, in press.

Grant, P. R. (1972*a*). Bill dimensions of the three species of *Zosterops* on Norfolk Island. *Syst. Zool.* **21**, 289–91.

Grant, P. R. (1972*b*). Convergent and divergent character displacement. *Biol. J. Linn. Soc.* **4**, 39–68.

Grant, P. R. (1977). Review of *Island Biology, Illustrated by the Land Birds of Jamaica* by David Lack. *Bird-Banding*, **48**, 296–300.

Grant, P. R. (1978). Recent evolution of *Zosterops lateralis* on Norfolk Island, Australia. *Can. J. Zool.* **56**, 1624–6.

Grant, P. R. (1979*a*). Evolution of the Chaffinch, *Fringilla coelebs*, on the Atlantic Islands. *Biol. J. Linn. Soc.* **11**, 301–32.

Grant, P. R. (1979*b*). Ecological and morphological variation of Canary Island blue tits *Parus caeruleus* (Aves: Paridae). *Biol. J. Linn. Soc.* **11**, 103–29.

Grant, P. R. (1981*a*). Speciation and the adaptive radiation of Darwin's finches. *Amer. Sci.* **69**, 653–63.

Grant, P. R. (1981*b*). The feeding of Darwin's finches on *Tribulus cistoides* (L.) seeds. *Anim. Behav.* **29**, 785–93.

Grant, P. R. (1981*c*). Patterns of growth in Darwin's finches. *Proc. Roy. Soc. Lond.* B, **212**, 403–32.

Grant, P. R. (1982). Variation in the size and shape of Darwin's finch eggs. *Auk*, **99**, 15-23.

Grant, P. R. (1983). The role of interspecific competition in the adaptive radiation of Darwin's Finches. In *Patterns of Evolution in Galápagos Organisms*, ed. A. Levinton and R. I. Bowman. Amer. Assoc. Adv. Sci. Special Publ. 1, Pacific Division, San Francisco.

Grant, P. R. and Abbott, I. (1980). Interspecific competition, island biogeography and null hypotheses. *Evolution*, **34**, 332–41.

Grant, P. R. and Boag, P. T. (1980). Rainfall on the Galápagos and the demography of Darwin's finches. *Auk*, **97**, 227–44.

Grant, P. R., Boag, P. T. and Schluter, D. (1979). A bill color polymorphism in young Darwin's Finches. *Auk*, **96**, 800–2.

Grant, P. R. and Grant, B. R. (1980*a*). The breeding and feeding characteristics of Darwin's finches on Isla Genovesa, Galápagos. *Ecol. Monogr.* **50**, 381–410.

Grant. P. R. and Grant, B. R. (1980*b*). Annual variation in finch numbers, foraging and food supply on Isla Daphne Major, Galápagos. *Oecologia*, **46**, 55–62.

Grant, P. R., Grant, B. R., Smith, J. N. M., Abbott, I. J. and Abbott, L. K. (1976). Darwin's finches: population variation and natural selection. *Proc. natl. Acad. Sci.* **73**, 257–61.

Grant, P. R. and Price, T. D. (1981). Population variation in continuously varying traits as an ecological genetics problem. *Amer. Zool.* **21**, 795–811.

Grant, P. R., Price, T. D. and Snell, H. (1980). The exploration of Isla Daphne Minor. *Notic. de Galáp.* **31**, 22–7.

Grant, P. R. and Schluter, D. (1983). Interspecific competition inferred from patterns of guild structure. In *Ecological Communities: Conceptual Issues and the Evidence*, ed. D. R. Strong, D. S. Simberloff and L. G Abele. Princeton Univ. Press, Princeton, New Jersey.

Grant, P. R., Smith, J. N. M., Grant, B. R., Abbott, I. J. and Abbott, L. K.

(1975). Finch numbers, owl predation and plant dispersal on Isla Daphne Major, Galápagos. *Oecologia*, **19**, 239–57.

Hamilton, T. H. and Rubinoff, I. (1967). On predicting insular variation in endemism and sympatry for the Darwin's finches in the Galápagos archipelago. *Am. Nat.* **101**, 161–71.

Hamilton, W. D. (1964). The genetical evolution of social behaviour. I, II. *J. Theor. Biol.* **7**, 1–52.

Harris, M. P. (1973). The Galápagos avifauna. *Condor*, **75**, 265–78.

Harris, M. P. (1974). *A Field Guide to the Birds of Galápagos.* Collins, London.

Horn, H. S. and May, R. M. (1977). Limits to similarity among coexisting competitors. *Nature*, **270**, 660–1.

Howard, R. and Moore, A. (1980). *A Complete Checklist of Birds of the World: Order Passeriformes.* Oxford University Press, Oxford.

Hutchinson, G. E. (1951). Copepodology for the ornithologist. *Ecology*, **32**, 571–7.

Hutchinson, G. E. (1957). Concluding remarks. *Cold Springs Harbor Symp. Quant. Biol.* **22**, 415–27.

Jackson, J. B. C. (1981). Interspecific competition and species' distributions: the ghosts of theories and data past. *Amer. Zool.* **21**, 889–901.

Jo, N. (1976). Karyotypic analysis of Darwin's Finches. Unpublished Master's thesis, San Francisco State University, San Francisco.

Lack, D. (1941). The evolution of the Galápagos finches. *Ibis*, **5**, Ser. 14, 637–8.

Lack, D. (1943). *The Life of the Robin.* M. F. and G. Witherby, London.

Lack, D. (1945). The ecology of closely related species with special reference to cormorant (*Phalacrocorax carbo*) and shag (*P. aristotelis*). *J. Anim. Ecol.* **14**, 12–16.

Lack, D. (1946). Competition for food by birds of prey. *J. Anim. Ecol.* **15**, 123–9.

Lack, D. (1949). The significance of ecological isolation. In *Genetics, Paleontology, and Evolution*, ed. G. L. Jepson, E. Mayr and G. G. Simpson, pp. 299–308. Princeton University Press, Princeton, New Jersey.

Lack, D. (1950). Breeding seasons in the Galápagos. *Ibis*, **92**, 268–78.

Lack, D. (1961). *Darwin's Finches.* Harper Torchbooks, New York.

Lack, D. (1965). Evolutionary ecology. *J. Anim. Ecol.* **34**, 223–31.

Lack, D. (1968). *Ecological Adaptations for Breeding in Birds.* Methuen. London.

Lack, D. (1969). Sub-species and sympatry in Darwin's finches. *Evolution*, **23**, 252–63.

Lack, D. (1971). *Ecological Isolation in Birds.* Harvard University Press, Cambridge, Mass.

Lack, D. (1973). My life as an amateur ornithologist. *Ibis*, **115**, 421–31.

Lack, D. (1976). *Island Biology. Illustrated by the Land Birds of Jamaica.* Blackwell Scientific, Oxford.

Lack, D. and Southern, H. N. (1949). The birds of Tenerife. *Ibis*, **91**, 607–26.

Lande, R. (1979). Quantitative genetic analysis of multivariate evolution, applied to brain: body size allometry. *Evolution*, **33**, 402–16.

Lanyon, W. E. (1978). Revision of the *Myiarchus* flycatchers of South America. *Bull. Amer. Mus. of Nat. Hist.* **161**, article 4, 427–628.

Laurie, A. (1982). Marine iguanas – where have all their babies gone? *Notic. de Galáp.* **35**, 17–19.

Levins, R. (1968). *Evolution in Changing Environments.* Princeton University Press, Princeton, New Jersey.

MacArthur, R. H. (1958). Population ecology of some warblers of northeastern coniferous forests. *Ecology*, **39**, 599–619.

MacArthur, R. H. (1972). *Geographical Ecology.* Harper and Row, New York.

MacArthur, R. H. and Wilson, E. O. (1967). *The Theory of Island Biogeography.* Princeton University Press, Princeton, New Jersey.

MacFarland, C. G. and Reeder, W. G. (1974). Cleaning symbiosis involving Galápagos tortoises and two species of Darwin's finches. *Zeits. fur Tierpsychol.* **34**, 464–83.

Marlow, R. W. and Patton, J. L. (1981). Biochemical relationships of the Galápagos Giant tortoises (*Geochelone elephantopus*). *J. Zool. Lond.* **195**, 413–22.

Mayr, E. (1966). *Animal Species and Evolution.* Harvard University Press, Cambridge, Mass.

McBirney, A. R. and Aoki, K. (1966). Petrology of the Galápagos Islands. In *The Galápagos, Proc. of the Symposia of the Galápagos Int. Scientific Project*, ed. R. I. Bowman. University of California Press, Berkeley.

Moore, T. (1980). *Galápagos – Islands Lost in Time.* Viking Press.

Murray, B. G. Jr (1982). Territorial behavior of the blue-black grassquit. *Condor*, **84**, 119.

Nuechterlein, G. L. (1981). Courtship behavior and reproductive isolation between Western Grebe color morphs. *Auk*, **98**, 335–49.

O'Donald, P., Wedd, N. S. and Davis, J. W. F. (1974). Mating preferences and sexual selection in the Arctic Skua. *Heredity*, **33**, 1–16.

Paepe, P. (1955). Geologie van Isla Daphne Major (Islas Galápagos Ecuador). *Natuurwet. Tijdschrift*, **48**, 67–80.

Park, T. (1939). Analytical population studies in relation to general ecology. *Amer. Midl. Nat.* **21**, 235–55.

Park, T., Gregg, E. V. and Lutherman, C. Z. (1941). Studies in population physiology. X. Interspecific competition in populations of granary beetle *Physiol. Zool.* **14**, 395–430.

Paterson, H. E. H. (1978). More evidence against speciation by reinforcement. *S. Afr. J. Sci.* **74**, 369–71.

Patton, J. L., Yang, S. Y. and Myers, P. (1975). Genetic and morphologic divergence among introduced rat populations (*Rattus rattus*) of the Galápagos archipelago, Ecuador. *Syst. Zool.* **24**, 296–310.

Paynter, R. A., Jr, ed. (1970). *Check-list of Birds of the World*. Vol. 13. Cambridge, Mass. Museum of Compar. Zool., Harvard University.

Pianka, E. R. (1974). *Evolutionary Ecology*. Harper and Row, New York.

Power, D. M. (1975). Similarity among avifaunas of the Galápagos Islands. *Ecology*, **56**, 616–26.

Price, T. D. and Millington, S. J. (1982). Birds on Daphne Major. *Notic. de Caláp*. **35**, 25–7.

Raikow, R. J. (1976). The origin and evolution of the Hawaiian Honeycreepers (Drepanididae). *Living Bird*, **15**, 95–117.

Ratcliffe, L. M. (1981). Species recognition in Darwin's Ground Finches (*Geospiza*, Gould). Unpublished Ph.D. thesis, McGill University, Montreal.

Ricklefs, R. E. and Cox, G. W. (1972). Taxon cycles in the West Indian avifauna. *Am. Nat.* **106**, 195–219.

Rothstein, S. I. (1973). Relative variation of avian morphological features: relation to the niche. *Am. Nat.* **107**, 796–9.

Rowher, S., Fretwell, S. D. and Niles, D. M. (1980). Delayed maturation in passerine plumage and the deceptive acquisition of resources. *Am. Nat.* **115**, 400–37.

Sammalisto, L. (1966). Censusing Darwin's finches. *Notic. de Caláp*. **7/8**, 13–17.

Schluter, D. (1982*a*). Distributions of Galápagos ground finches along an altitudinal gradient; the importance of food supply. *Ecology*, **63**, 1504–17.

Schluter, D. (1982*b*). Seed and patch selection by Galápagos ground finches: relation to foraging efficiency and food supply. *Ecology*, **63**, 1106–20.

Schluter, D. and Grant, P. R. (1982). The distribution of *Geospiza difficilis* in relation to *G. fuliginosa* in the Galápagos islands: tests of three hypotheses. *Evolution*, **36**, 1213–1226.

Simberloff, D. and Boecklen, W. (1981). Santa Rosalia reconsidered: size ratios and competition. *Evolution*, **35**, 1206–28.

Simberloff, D. and Connor, E. F. (1981). Missing species combinations. *Am. Nat.* **118**, 215–39.

Smith, J. N. M., Grant, P. R., Grant, B. R., Abbott, I. J. and Abbott, L. K. (1978). Seasonal variation in feeding habits of Darwin's Ground Finches. *Ecology*, **59**, 1137–50.

Smith, J. N. M. and Sweatman, H. P. (1976). Feeding habits and morphological variation in Cocos Finches. *Condor*, **78**, 244–8.

Snow, D. (1966). Moult and the breeding cycle in Darwin's finches. *J. Orn.* **107**, 283–91.

Steadman, D. W. (1981). Vertebrate fossils in lava tubes in the Galápagos Islands. In B. F. Beck (ed.) *Proc. Eighth Int. Congr. of Speleology*, pp. 549–56.

Steadman, D. W. (1982). The origin of Darwin's finches (Fringillidae, Passeriformes). *Trans. San Diego Soc. Nat. Hist.* **19**, 279–96.

Strong, D. R. Jr, Szyska, L. A. and Simberloff, D. S. (1979). Tests of community-wide character displacement against null hypotheses. *Evolution*, **33**, 897–913.

Sulloway, F. J. (1982*a*). Darwin and his finches: the evolution of a legend. *J. Hist. Biol.* **15**, 1–53.

Sulloway, F. J. (1982*b*). The *Beagle* collections of Darwin's Finches (Geospizinae). *Bull. Brit. Mus. nat. Hist.* (*Zool.*), **43** (2), 49–94.

Templeton, A. (1981). Mechanisms of speciation – a population genetic approach. *Ann. Rev. Ecol. Syst.* **12**, 23–48.

Thielcke, G. (1973). On the origin of divergence of learned signals (songs) in isolated populations. *Ibis*, **115**, 511–16.

Thornton, I. (1971). *Darwin's Islands. A Natural History of the Galápagos.* Natural History Press, Garden City, New York.

Van Valen, L. (1965). Morphological variation and width of ecological niche. *Am. Nat.* **99**, 377–90.

Waage, J. K. (1979). Reproductive character displacement in *Calopteryx* (Odonata: Calopterygidae). *Evolution*, **33**, 104–16.

Walters, M. (1980). *The Complete Birds of the World.* David and Charles, London.

White, M. J. D. (1978). *Modes of Speciation.* W. H. Freeman, San Francisco.

Wiens, J. A. (1977). On competition and variable environments. *Amer. Sci.* **65**, 590–7.

Wiggins, I. L. and Porter, D. M. (1971). *Flora of the Galápagos Islands.* Stanford University Press, Stanford, California.

Williams, H. (1966). Geology of the Galápagos Islands. In *The Galápagos. Proc. of the Symposia of the Galápagos Int. Scientific Project*, ed. R. I. Bowman. University of California Press, Berkeley.

Williamson, M. (1981). *Island Populations.* Oxford University Press, Oxford.

Williamson, P. G. (1981). Morphological stasis and developmental constraints: real problems for neo-Darwinism. *Nature*, **294**, 214–15.

Wright, S. (1980). Genic and organismic selection. *Evolution*, **34**, 825–43.

Yang, S. Y. and Patton, J. L. (1981). Genic variability and differentiation in the Galápagos finches. *Auk*, **98**, 230–42.

Zann, R. (1976). Inter- and intraspecific variation in the courtship of three species of Grassfinches of the subgenus *Poephila* (Gould) (Estrildidae). *Zeits. fur Tierpsychol.* **41**, 409–33.

PART ONE: DESCRIPTION

Chapter I: GALÁPAGOS SCENE

The country was compared to what we might imagine the cultivated parts of the Infernal regions to be.

Charles Darwin: MS. Diary of the Voyage of H.M.S. *Beagle*

THE ISLANDS

On the evening of 10 December 1938 a squalid uncomfortable Ecuadorean trading schooner carried our expedition down the Guayas river and out into the Pacific. Two days later a marked drop in temperature and an abundance of sea birds proclaimed the equatorial cold of the Humboldt current, and before dawn on 14 December anchor was dropped in Wreck Bay, Chatham, the most south-easterly island of the Galapagos. We emerged at daybreak to see a drab, short-tailed finch hopping about the deck, our first uninspiring view of one of Darwin's finches.

The numerous recent travel books decribing the Galapagos—the 'Enchanted Isles'—had not sufficiently prepared us for the inglorious panorama. Behind a dilapidated pier and ramshackle huts stretched miles of dreary, greyish brown thornbush, in most parts dense, but sparser where there had been a more recent lava flow, and the ground still resembled a slag heap. The land rose gradually, with no exciting features, to a sordid cultivated region, beyond which, partly concealed in cloud, were green downs, the only refreshing spot in the scene.

Closer acquaintance in the next four months only increased the initial depression. The Galapagos are interesting, but scarcely a residential paradise. The biological peculiarities are offset by an enervating climate, monotonous scenery, dense thorn scrub, cactus spines, loose sharp lava, food deficiencies, water shortage, black rats, fleas, jiggers, ants, mosquitoes, scorpions, Ecuadorean Indians of doubtful honesty, and dejected, disillusioned European settlers. Admittedly these are merely discomforts, but their effect is cumulative in a shut-in tropical island. Moreover, for one of the party there was serious trouble, dysentery. Perhaps our souls were 'made of lead, too dull, too ponderous, to mount up to the incomprehensible glory that Travel lifts men to'. But Old Fortunatus had not visited the Enchanted Isles. Charles Darwin, who had, knew better. 'No doubt it is a high satisfaction to behold various countries and the many races of mankind, but the pleasures gained at the time do not counterbalance the evils. It is necessary to look forward to a harvest, however distant that may be, when some fruit will be reaped, some good effected.'

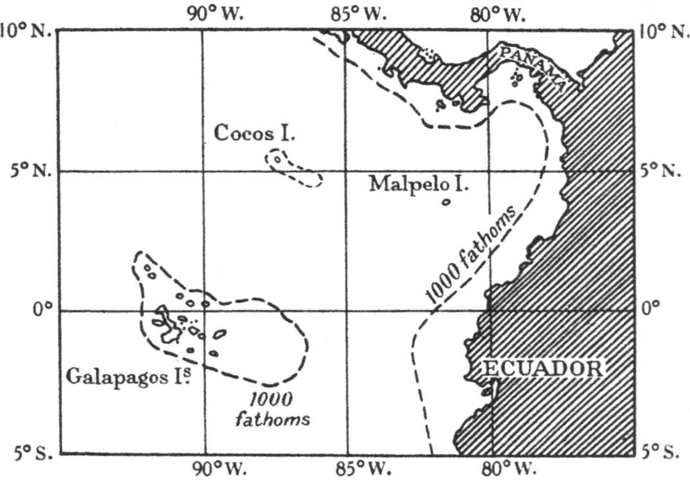

Fig. 1. Position of Galapagos Islands.

The Galapagos lie on the Equator in the eastern Pacific, 600 miles from Ecuador and 1000 from Panama. The only adjacent islands are forested Cocos and the rock Malpelo, both about 600 miles to the north-east. West of the Galapagos, there is no land of any sort for 3000 miles.

The largest Galapagos island is Albemarle, about 80 miles long with a highest point rather over 4000 ft. above the sea. Several of the other islands are between 10 and 20 miles across and rise to a height of 2000 or 3000 ft., while there are many other smaller and low-lying islands. The islands are volcanic in origin, and volcanic activity still occurs on some of them. Where there has been recent activity, the ground is jagged and the contours are irregular, but on the older islands, notably much of Chatham, the ground is smooth and the contours are gently rounded. There are circular crater lakes on Albemarle and Chatham, the latter being drained by a permanent stream, and there are isolated springs on Charles, James and Albemarle, and slightly brackish wells on Indefatigable. Otherwise there is no permanent fresh water, but during the rainy season temporary—often extremely temporary—pools and streams are formed, which drain away rapidly through the lava. Immediately after a cloudburst on Indefatigable a previously dry stream-bed contained an. almost impassable torrent, but it could again be crossed dryshod a few hours later.

The rainy season lasts from about mid-December to early March, but it is variable, in some years continuing into June, while in other years hardly any rain falls. It is a period of calms and bright sunshine interrupted by thunder, squalls, light showers and heavy cloudbursts. During the rest of the year the south-east trade wind blows, and normally no rain falls on the coast, but a thick mist, the garua, covers the higher ground. The day temperatures are the coolest of any equatorial region in the world, due to the Humboldt current, an upwelling of cold water against the west coast of South America.

The shores of the Galapagos chiefly consist of low cliffs and black lava boulders, ornamented with scarlet crabs and black archaic iguanas, while locally there are white sand beaches with sea-lions and turtles, occasional dense mangrove swamps, and a few tall cliffs.

The most prominent objects in the barren and arid lowlands are the tall tree-cactuses, the dildo trees and torch thistles of the buccaneers. The long, branched, fluted cylinders of *Cereus* and the fleshy, spine-covered pads of the prickly pear *Opuntia* are both carried to about 30 ft. above the ground. The other vegetation consists of more ordinary bushes and trees, some extremely thorny, such as *Acacia*, others not, such as *Bursera*, *Croton*, *Maytenus* and the poisonous manzanilla *Hippomane mancinella*. Where lava flows have occurred recently, much of the ground is bare, and progress is easy except where the surface is too jagged or loose, or where there are deep clefts. At the other extreme, there are many parts where the bushes form a dense tangle through which it is impossible to force a way, even with a machete. Elsewhere, the vegetation is moderately open. For most of the year this lowland zone looks parched and dry, but for a brief period immediately following the heavy rains it appears bright green and deceptively fertile.

Inland and high up, the scene is strikingly different—humid forest with rich black soil and tall trees covered with ferns, orchids, lichens and mosses. One of the most characteristic trees is *Scalesia pedunculata*, a member of the Compositae (daisy family); others include species of *Psidium*, *Pisonia* and *Zanthoxylum*, and locally there are groves of tree ferns. This rich growth is possible because not only does much more rain fall on this region than on the coast, but outside the rainy season it is kept damp by the thick mist. Humid forest is found only on the larger and higher islands, James, Indefatigable, Albemarle, and Charles, with a small area on Abingdon, while it may formerly have occurred on Chatham. On Chatham the whole, and on Albemarle, Indefatigable and Charles a large part, of this zone is now under cultivation, chiefly with coffee, sugar and the common tropical fruits.

The height above sea-level at which the humid forest is found varies considerably with local conditions. It commences at a lower altitude on the windward (south-east) than the leeward side of each island, and is also affected by the period of the last volcanic activity and by other factors. Under favourable conditions such forest is found within 500 or 600 ft. of sea-level, but with more arid conditions it fails altogether, and cactus and thorn scrub may then extend to the mountain tops.

Between the arid lowlands and the fertile humid forest, there is a region where the vegetation is transitional in type. Here the thorn and cactus become progressively scarcer and the forest trees with epiphytic lichens progressively commoner, the farther that one travels inland from the coast. On the larger islands, extensive areas are covered with vegetation of this transitional type.

In the highest parts of the humid forest the trees decrease in height and bear a particularly luxuriant cover of epiphytic ferns, orchids and mosses. Then they give place to open country with grass, ferns, bracken, clubmosses, liverworts, mosses and occasional bushy thickets. Extensive open country is found on the three highest islands, Indefatigable, Albemarle and Chatham, and a small area on Charles. In this zone on Indefatigable, the curious traveller can walk on the grassy margins of extinct craters, or tread a continuous carpet of lush liverworts, or rest in a miniature Coal Measure forest of erect 3 ft. clubmosses; while if a cloudburst occurs, he can watch racing streams which suddenly fall far out of sight into deep, narrow fissures, or can shelter in a fern-festooned cave formed of a huge burst bubble of lava. On the other islands, the open uplands have been much changed by the grazing of introduced cattle, and grasses unduly predominate, while on Chatham tropical guava has run wild and locally forms dense areas of scrub. The downs of Chatham, with their thick mists, rounded contours and herds of cattle and horses, look rather like Dartmoor, but brilliantly green, with ponds pink with the floating plant azolla, and frigate birds soaring over a crater lake. The top of Albemarle, where volcanic activity has been more recent, is said to be more barren than either Chatham or Indefatigable.

All the Galapagos islands look much alike, and the chief variations are due to differences in altitude and in the time of the last volcanic activity. But Cocos, where one of Darwin's finches occurs, looks quite different. Lying some 600 miles north-east of the Galapagos and some 300 miles from Panama, Cocos is a typical tropical island, with a warm, moist climate and rich forest coming down to the sea. It belongs to Costa Rica, and attracts treasure hunters.

The Galapagos, as is well known, received their name from the giant land tortoises, which provided much food, first for the buccaneers, then for the whalers. Later these animals were heavily exploited for oil, and now the final remnants are being eaten by the settlers. The peculiar land iguana, some 3 ft. long, is also nearly extinct, but the even more curious marine iguana is still numerous. The only conspicuous land mammals are introduced forms which have run wild, cattle, horses, donkeys, pigs, dogs, goats and black rats, and these animals are having a harmful effect on the native plants and animals. The smaller native land mammals, the birds and the reptiles will be mentioned later, in Chapter XII. Along the shore, sea-lions are common and friendly, but these and the conspicuous sea birds are outside the scope of this book.

This brief account shows the background, sometimes curious

and nearly always unpleasant, in which the field work on Darwin's finches was carried out. Fuller accounts of the islands, frequently misleading, occur in many travel books, a valuable survey of the plant life is given by Stewart (1915), while the most careful and accurate general description is still that of the naturalist of H.M.S. *Beagle*.

HUMAN HISTORY

The human history of the Galapagos has been surveyed by Rose (1924). It is mainly a tale of disaster, tempered by squalid crime. The islands were first made known by Fray Tomás de Berlanga, bishop of Panama, who came in 1535 and like many after him experienced great privations, searched vainly for water, and chewed cactus: 'It looked as though God had caused it to rain stones.' The Elizabethan Sir Richard Hawkins briefly commented: 'They are desert, and beare no fruite.' The buccaneers Cowley, Dampier and Wafer misled subsequent visitors by recording green vegetation, water and even rivers in June 1684; evidently the year in which the *Bachelors' Delight* called must have had an exceptionally prolonged rainy season. Some years later another buccaneer, Woodes Rogers, took Alexander Selkirk (the original Robinson Crusoe) off Juan Fernandez, sacked Guayaquil, and came thence to the Galapagos, where he found 'nothing but loose Rocks, like Cynders, very rotten and heavy ...tho' there is much shrubby Wood and some Greens, yet there's not the least Sign of Water, nor is it possible that any can be contained on such a Surface'.

After the buccaneers came the whalers, and then the first scientific expedition. By this time the devastation of rural England by the Industrial Revolution had provided a parallel with natural vulcanism, and in 1835 Charles Darwin could compare the Galapagos scene with the iron foundries of Staffordshire or the cultivated parts of the Infernal regions. Likewise Captain FitzRoy recorded 'black dismal-looking heaps of broken lava, forming a shore fit for Pandemonium'.

Until 1832 the Galapagos belonged to Spain and remained uninhabited, except temporarily by Patrick Watkins, a drunken Irishman. But after the War of Liberation the islands were claimed by Ecuador. Water having by now been found in the

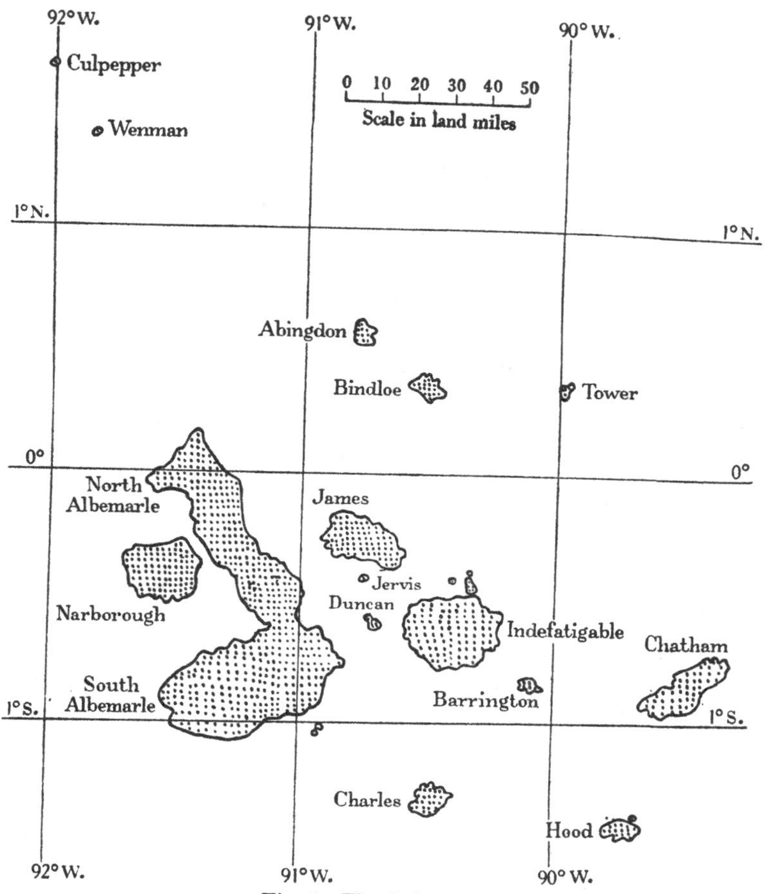

Fig. 2 The Galapagos.

TABLE I. NAMES OF THE GALAPAGOS ISLANDS

English name	Ecuadorean name
Galápagos	Archipielago de Colón
Abingdon	Pinta
Bindloe	Marchena
Tower	Genovesa
James	Santiago
Jervis	Rábida
Indefatigable	Santa Cruz (also Chavez)
Duncan	Pinzón
Albemarle	Isabela
Narborough	Fernandina
Barrington	Santa Fé
Chatham	San Cristobal
Hood	Española
Charles	Santa Maria (also Floreana)

Note. The names of the other islands, including Culpepper and Wenman, were not changed by the Ecuadoreans.

hills, a settlement was made on Charles, where several hundred persons lived for a time, but the colony was eventually abandoned. Charles was again settled in 1893, but four years later the colonists transferred to South Albemarle, where a large settlement still persists. Charles then remained uninhabited until the nineteen-twenties, when a few Europeans arrived, and though these have since died, some notoriously, others have taken their place. About the same time, Indefatigable was settled by Europeans, and though many of them later left or died, a small colony has remained. The only other island to be permanently settled is Chatham, where a large colony was founded in the latter part of the nineteenth century and has since persisted, despite occasional attempts, one successful, to murder the governor. Hence up to 1939 four Galapagos islands have been settled; but the establishment of U.S. bases in 1942 has yet to be recorded. Our own party stayed mainly on Chatham and Indefatigable, while brief visits were paid to other islands, including Hood and Tower.

In 1892 the four hundredth anniversary of the discovery of America, the Ecuadorean Government renamed the Galapagos archipelago after Columbus, who never went there, and changed the traditional names of the islands, the new names mostly being those of men associated with Columbus. These new names have been accepted in official publications. It is highly desirable to have a standardized system of place-names, and if this had been a geographical treatise, the Ecuadorean names would have been used unhesitatingly. But the English names of the islands have been used throughout the biological literature, so that, after much hesitation, they have been retained here. They were mostly given in 1684 by the patriotic buccaneers of the *Bachelors' Delight*, which explains the significance of Charles, James and the like, while a few others were named by Captain Colnett, who surveyed the islands in 1793. The modern names of the islands, together with their English equivalents, are given in Table I on the previous page.

THE FINCHES

Captain Colnett was the first to comment, somewhat injudiciously, on the now famous finches. 'This island contains no great number, or variety, of land birds, and those I saw were not remarkable for their novelty or beauty.' He described small birds 'resembling the Java sparrow, in shape and size, but of black plumage'. Captain Porter of the U.S. frigate *Essex*, who captured twelve British whaling ships in Galapagos waters during the war of 1812, also mentioned 'a small black bird, with a remarkably short and strong bill, and a shrill note'.

Then came the visit of H.M.S. *Beagle* in 1835. The bird specimens collected by Darwin were described by Gould (1837), who recognized the finches as an entirely new group, while the influence which these and other Galapagos animals exerted on Darwin's views on evolution is common knowledge. In his *Evolutionary Notebook*, started in 1837, Darwin wrote: 'In July opened first note-Book on "Transmutation of Species". Had been greatly struck from about month of previous March on character of S. American fossils—& species on Galapagos Archipelago.—These facts origin (especially latter) of all my views' (Barlow, 1933). Again, in the first private announcement of his new views, a letter to Joseph Hooker on 11 January 1844, Darwin wrote: 'I was so struck with the distribution of the Galapagos organisms, &c. &c., and with the character of the American fossil mammifers &c., &c., that I determined to collect blindly every sort of fact, which could bear any way on what are species....At last gleams of light have come, and I am almost convinced (quite contrary to the opinion I started with) that species are not (it is like confessing a murder) immutable' (F. Darwin, 1887).

Darwin first questioned the mutability of species when actually in the Galapagos, through finding different forms of the mocking-bird and tortoise on different islands (Barlow, 1935). The finches, with several species on each island, are more complex, and their influence was apparently retrospective. Thus, in Darwin's private diary of the voyage, the finches are not mentioned (Barlow, 1933), and even in the first published edition of the *Journal*, in 1839, they receive only brief notice, without particular comment. However, this paragraph was considerably amplified in the second edition of 1845: 'The remaining land-birds form a most singular group of finches, related to each other in the structure of their beaks, short tails, form of body and plumage. All these species are peculiar to this archipelago.' Darwin went on to describe 'the perfect gradation in the size of the beaks in the different species', and concluded: 'Seeing this gradation and diversity of structure in one small, intimately related group of birds, one might really fancy that from an original paucity of birds in this archipelago, one species had been taken and modified for different ends.' This last phrase is the most significant in the whole book, and is Darwin's first public pronouncement on a subject the elaboration and generalization of which was to occupy the next fifteen years of his life.

The next collection of Darwin's finches was made by Habel in 1868, and the birds were described by Salvin in 1876. These specimens, together with those collected by Darwin, are now in the British Museum of Natural History. In 1897 Ridgway described two further large collections, one made by the naturalists of the *Albatross* in 1888, and the other by Baur in 1891. The former collection is in the United States National Museum at Washington, while most of Baur's specimens passed to the Rothschild collection, formerly at Tring and now in the American Museum of Natural History, New York. The Rothschild collection also received the specimens of the Webster-Harris expedition a few years later, and Rothschild and Hartert wrote papers on the birds in 1899 and 1902. In 1904 Snodgrass and Heller described a large collection made by the Hopkins-Stanford expedition. Finally, much the biggest collection was made by the expedition of the California Academy of Sciences in 1905–6, this being ably monographed by Swarth in 1931. As a result of all these visits, Darwin's finches are more adequately represented by museum specimens than almost any other group of birds.

Some of the collectors, notably Gifford (1919), have published field notes on the habits of the finches, particularly their feeding habits. Much further information on ecology and breeding behaviour was obtained by our expedition, which also brought back thirty living birds of four species. These were originally intended for the Zoological Society of London, but travelled so badly that they were taken instead to the California Academy of Sciences, San Francisco, where they remain. At the latter institute I also worked on the extensive skin collections of Darwin's finches, and later studied all the other large collections referred to above.

Darwin's finches are dull to look at, not only in their orderly ranks in museum trays, but also when they hop about the ground or perch in the trees of the Galapagos, making dull unmusical noises. Only the variety of their beaks and the number of their species excite attention—small finch-like beaks, huge finch-like beaks, parrot-like beaks, straight wood-boring beaks, decurved flower-probing beaks, slender warbler-like beaks; species which look very different and species which look closely similar. 'How astonishing', wrote Sydney Smith, 'are the freaks and fancies of nature! To what purpose, we say, is a bird placed in the forests of Cayenne, with a bill a yard long, making a noise like a puppy dog, and laying eggs in hollow trees? The toucan, to be sure, might retort—to what purpose were gentlemen in Bond Street created?...There is no end to such questions. So we will not enter into the metaphysics of the toucan.' But such amiable and ineffective speculation on the purposes of nature and the ways of birds was soon to give place to a careful analysis which drastically changed man's basic conception of his place in the world, the change initiated by Darwin, when he wrote of the Galapagos: 'The natural history of these islands is eminently curious, and well deserves attention. Most of the organic productions are aboriginal creations, found nowhere else; there is even a difference between the inhabitants of the different islands; yet all show a marked relationship with those of America, though separated from that continent by an open space of ocean, between five and six hundred miles in width. The archipelago is a little world within itself, or rather a satellite attached to America, whence it has derived a few stray colonists and has received the general character of its indigenous productions. Considering the small size of these islands, we feel the more astonished at the number of their aboriginal beings, and at their confined range. Seeing every height crowned with its crater, and the boundaries of most of the lava-streams still distinct, we are led to believe that within a period geologically recent the unbroken ocean was here spread out. Hence, both in space and time, we seem to be brought somewhat near to that great fact—that mystery of mysteries—the first appearance of new beings on this earth.'

Chapter II: CLASSIFICATION

Something more is included in our classification than mere resemblance.... Propinquity of descent,—the only known cause of the similarity of organic beings, —is the bond, hidden as it is by various degrees of modification, which is partially revealed to us.... Thus the grand fact in natural history of the subordination of group under group, which, from its familiarity does not sufficiently strike us, is in my judgement explained.

Charles Darwin: *The Origin of Species*, Ch. XIII[1]

DIFFICULTIES

Before any discussion of them is possible, Darwin's finches must be named, although this in itself commits one to a partial interpretation of the problems of their evolution. Fortunately, there is no longer any serious dispute as to the correct naming of the various forms. At one time matters were otherwise, and authorities differed considerably as to which specimens should be ascribed to different species, which were island forms of the same species, and which were merely varieties.

These differences of opinion arose from the inadequacy of the collected material, but they were considerably accentuated by the unfamiliar nature of the variations found among the finches, which often ran contrary to the experience of museum systematists accustomed to working with European or North American birds. In continental passerine birds, closely related species tend to differ from each other chiefly in plumage, and they are usually similar in beak and other structural characters. Differences in beak more usually characterize the broader units, the different genera. In Darwin's finches, on the other hand, closely related species usually differ markedly in beak, but little, if at all, in plumage, and plumage differences are chiefly important in distinguishing the different genera.

A further unusual feature in Darwin's finches is that some of the species are highly variable. For example, individuals of the ground-finch *Geospiza fortis* are so variable in beak that they were for a long time considered to belong to at least two, and by some

[1] To provide a succinct quotation, I have taken the last sentence of the above from a paragraph which precedes the rest. But there is no distortion of Darwin's meaning.

authorities to three or more, separate species. Extensive collections were needed to establish that in fact only one exceptionally variable form was involved. Had only two specimens of this species been known, one at the minimum of size and the other at the maximum, systematists might well have placed them in separate genera.

ENGLISH NAMES

The species of Darwin's finches are most clearly and simply referred to by their scientific Latin names, but to help the reader to picture the birds, I have given each species a brief descriptive English designation. English names have been invented for some of the species by previous systematists, and Hellmayr (1938) has produced a complete list. But Hellmayr's names tend to be mere translations of the Latin ones, and are not sufficiently descriptive to help the general reader, while a few are actually inappropriate. Because of these objections, and because Hellmayr's names are mostly of recent origin and are not yet sanctioned by custom, I considered it justifiable to replace them by a new and more appropriate set. This change should not cause confusion, as each species is always referred to by its scientific Latin name, and the preceding English name should be regarded merely as a supplementary description. There are not, of course, any local or traditional names for the finches in the Galapagos.

The genera, species and island forms of Darwin's finches are briefly summarized in Tables II, III and IV which follow later in this chapter. In addition, the species are illustrated in Fig. 3 (p. 19) and also in Plates I, IV, V and VIII. It is hoped that these illustrations, the three tables and the English designations of the birds will keep the characteristics of the various forms sufficiently in readers' minds for the argument of later chapters to be clear.

THE SUBFAMILY

On anatomical grounds, Snodgrass (1903), Sushkin (1925, 1929) and Lowe (1936) are agreed that the varied forms comprising Darwin's finches are all closely related to each other, including the peculiar warbler-finch *Certhidea*, which superficially looks much more like a warbler than a finch. *Certhidea* was placed with the other finches when first described by Gould in 1837, though

Darwin wondered whether this was correct. Later workers considered that the bird was not a finch at all, and placed it in various families, latterly with the American warblers (Mniotiltidae). However, the findings of Snodgrass have now been generally accepted, and our own field study has revealed strong similarities in breeding and other behaviour, confirming the close relationship of all the forms with each other, including *Certhidea*.

The whole group may be termed the Geospizinae, a subfamily of the finches (Fringillidae), to which Sushkin and Lowe showed that they are related. For simplicity they are referred to in this book as Darwin's finches. The term 'Galapagos finch' is less satisfactory, since one species, namely *Pinaroloxias inornata*, occurs not in the Galapagos, but on Cocos Island, 600 miles to the north-east. Similarities in plumage and anatomy show that *Pinaroloxias* is undoubtedly one of the Geospizinae. With this exception, Darwin's finches are confined to the Galapagos.

All of Darwin's finches are greyish brown, short-tailed birds, with fluffy rump feathers. In some species the two sexes are alike in plumage, in others the males are distinguished by a varying amount of black feathering, while the male *Certhidea* has an orange-tawny throat. All the species build roofed nests, large for the size of the bird, and lay white eggs spotted with pink, four to a clutch. All are territorial and monogamous, and in all species courtship includes display with nest material and the feeding of the female by the male. There are many other similarities in breeding behaviour, and many resemblances in internal anatomy. They vary in size from a small warbler to a very large sparrow, and their feeding habits are exceedingly diverse, as are their beaks, which range from delicate and thin to stout and huge.

GENERA

Swarth (1931), the latest authority, divided Darwin's finches into six genera, while a seventh has been used by some other workers. The beak differences between these seven subgroups are so marked that, if they were continental passerine birds, they would unhesitatingly be classified as separate genera. However, in most other respects, including plumage and breeding habits, some of these subgroups are closely similar to each other. Further, since Darwin's finches are in this book divided into only fourteen species, the employment of six or seven generic names seems excessive. For these reasons only four genera are used here, as set out in Table II.

TABLE II. GENERA OF DARWIN'S FINCHES

Genus	Designation	Full male plumage	Beak	Chief food	No. of species
Geospiza	Ground-finch	Wholly black	Finch-like (one longer)	Seeds (cactus in one case)	6
Camarhynchus	Tree-finch	Partly black or without black	Thick, shape variable	Insects (fruit in one case)	6
Certhidea	Warbler-finch	Orange-tawny throat	Slender, warbler-like	Small insects	1
Pinaroloxias	Cocos-finch	Wholly black	Slender decurved	Small insects	1

Note. As here used, *Geospiza* includes the species *scandens*, sometimes placed in a separate genus *Cactornis*, while *Camarhynchus* includes the three genera *Platyspiza*, *Camarhynchus* and *Cactospiza* of Swarth (1931).

DETERMINATION OF SPECIES AND SUBSPECIES

The scientific classification of birds, started by Aristotle and continued after a long interval by Gesner, Ray, Linnaeus and many others, was in the first place a tabulation of existing knowledge regarding the birds of their own countries. The species concept of the early naturalists, though at first implicit rather than explicit, was more or less sound, because it was based on familiar animals. It was early appreciated that birds consist of separate kinds or species, that the members of each kind have in common certain characteristics in which they differ from all other kinds, and that the individuals of one kind do not usually breed with those of any other.

Later, when travellers brought back bird specimens from distant parts, naturalists began to give names to birds which they had not studied in their natural haunts, a practice which has continued up to the present time. Darwin's finches, for example, have been named and described in seven big systematic papers, those of Gould (1841), Salvin (1876), Ridgway (1897), Rothschild and Hartert (1899, 1902), Snodgrass and Heller (1904) and Swarth (1931), while these and other authors, including Hellmayr (1938), have also contributed shorter papers and notes. But of these men, only Snodgrass and Heller had seen alive the birds of which they wrote, though Swarth was able to visit the Galapagos some years after his paper was published.

The naming of the first bird species was based on a wide experience of living birds in their natural homes. Obviously the determination of species from specimens brought by others from strange places is a much more superficial procedure. However, the great majority of the birds of the world have been named solely from the appearance of their skins, the latter preferably being collected as a series, together with information as to age, sex and locality. Despite the superficiality of this information, later field study has nearly always confirmed the diagnoses made by the skilled museum systematist, and has shown that the species which he has named are in fact separate breeding-units. It has sometimes been claimed that the museum worker is engaged in compiling a purely artificial catalogue of the animal kingdom. But, in birds at least, this statement is untrue, and the discontinuities which separate names imply exist in nature. Even in the exceptionally difficult group of Darwin's finches, our field observations have shown that no fundamental alterations are required in the classification by Swarth (1931) based purely on museum specimens, and the only changes introduced here have been with the object of reducing the number of essential names to a minimum. Owing to the comparative ease with which bird species can be distinguished, and helped by the fact that there are a moderately large but not an excessive number of species, avian systematics is in a far more advanced state than that of any other class of animals. As a result, birds provide particularly good material for studying the origin of species.

Widely distributed bird species tend to be subdivided into geographical races or subspecies. The term subspecies has been used rather loosely in some other groups of animals, so it may be emphasized that in birds it is synonymous with a geographical race. In continental birds, there is usually no difficulty in deciding whether two different forms should be classified as subspecies or species. Subspecies, as the term implies, differ from each other to a smaller extent than do full species, the differences chiefly involving shade of plumage and size. But a more important criterion is that of geographical distribution. Subspecies of the same species always breed in separate geographical regions, and where their respective breeding zones adjoin, they often interbreed freely and intergrade in characters. On the other hand, two forms which breed in the same region without normally interbreeding are always classified as separate species, however similar they may be to each other in appearance.

Since, as discussed later, new species evolve from subspecies, there is no absolute division between the two categories, and the decision in border-line cases is inevitably arbitrary. Difficulty occurs chiefly in regard to related forms which occupy separate geographical regions, like subspecies, but which differ from each other more markedly than is usual among races of the same species. This happens not uncommonly among related island forms, and as these are isolated from each other geographically, it is often impossible to know whether or not they would freely interbreed with each other if they met. In general, it is more useful to emphasize the affinities rather than the distinctiveness of such island forms. Hence in Darwin's finches I have wherever possible treated them as subspecies of one species rather than as separate species. This also has the advantage of reducing the number of essential names. On the other hand, Swarth treated many of these distinctive island forms as separate species. On such a point, nomenclature is arbitrary.

SPECIES OF DARWIN'S FINCHES

With the above points in mind, Darwin's finches have in this book been divided into fourteen species. The characteristics of these birds are considered in detail in later chapters. The chief differences concern the size and shape of the beak, and at this stage a sufficient summary is provided by Table III, together with Fig. 3, as set out overleaf.

ISLAND SUBSPECIES

Most species of Darwin's finches occur on a number of islands. In some cases the island populations differ sufficiently to justify division into subspecies, in other cases the differences are less marked, and in yet others they are barely perceptible. Provided the degree of difference is described, the question of which island forms should be named is unimportant and inevitably somewhat arbitrary. In this book I have treated a form as a separate race or subspecies when I could distinguish at least 75 per cent of the available specimens from the specimens of any other form. As a result, I have used fewer subspecific names than did Swarth

(1931). Incidentally, the general reader need not trouble to remember the subspecific names, since for most purposes each form is sufficiently indicated by its specific name together with the name of the island or islands on which it occurs. The distribution of the various species and island forms is shown in Table IV, p. 20.

TABLE III. SPECIES OF DARWIN'S FINCHES

	Scientific name	Descriptive designation
1.	*Geospiza magnirostris* Gould	Large ground-finch
2.	*Geospiza fortis* Gould	Medium ground-finch
3.	*Geospiza fuliginosa* Gould	Small ground-finch
4.	*Geospiza difficilis* Sharpe	Sharp-beaked ground-finch
5.	*Geospiza scandens* (Gould)	Cactus ground-finch
6.	*Geospiza conirostris* Ridgway	Large cactus ground-finch
7.	*Camarhynchus crassirostris* Gould	Vegetarian tree-finch
8.	*Camarhynchus psittacula* Gould	Large insectivorous tree-finch
9.	*Camarhynchus pauper* Ridgway	Large insectivorous tree-finch on Charles
10.	*Camarhynchus parvulus* (Gould)	Small insectivorous tree-finch
11.	*Camarhynchus pallidus* (Sclater and Salvin)	Woodpecker-finch
12.	*Camarhynchus heliobates* (Snodgrass and Heller)	Mangrove-finch
13.	*Certhidea olivacea* Gould	Warbler-finch
14.	*Pinaroloxias inornata* (Gould)	Cocos-finch

Notes. (i) *Geospiza conirostris* has obvious affinities with *G. scandens* and replaces it geographically, but it is so distinctive that it is given a separate specific name. It should perhaps be reckoned as part of the *G. scandens* superspecies, but this is not certain.

(ii) The distinctive island forms related to *Camarhynchus psittacula* have hitherto been regarded as separate species, but they are here united under *C. psittacula*, except for *C. pauper*, which, though no more distinctive than the rest, must be treated as a separate species, as it occurs together with *C. psittacula* on Charles without intermingling. The significance of this case is discussed later.

(iii) Since the genera *Certhidea* and *Pinaroloxias* each include only one species, there is no need to give both generic and specific names in the text, and the latter are normally omitted. But the names of the other species are required.

(iv) In *Camarhynchus*, confusion is possible as regards the gender of the specific names. The International Rules of Nomenclature state that where the specific name is an adjective it should follow the gender of the generic name, but where it is a noun in apposition it should keep its own gender. In 1837 Gould named *Camarhynchus psittacula*; *psittacula* is to be regarded as a noun in apposition, so Gould's alteration to *psittaculus* in 1841 has not been adopted. But when the species originally named *Geospiza parvula* and *Cactornis pallida* are, as here, transferred to the genus *Camarhynchus*, they become *parvulus* and *pallidus* respectively, as the latter are adjectives. There are an increasing number of systematists who feel that the termination of an adjectival specific name should not be changed in this way, particularly as so many modern generic names are indeterminate as to gender. This view has much to recommend it, but as the international rules of nomenclature are explicit on this point, and have not, as yet, been changed, they are followed in this book.

Fig. 3. Darwin's finches; male and female of each species. Numbers refer to Table III opposite. About $\frac{2}{7}$ life-size.

Table IV. The Distribution and Island Forms of Darwin's Finches

Species	Culpepper	Wenman	Tower	Abingdon	Bindloe	James	Indefatigable	Albemarle	Narborough	Barrington	Chatham	Hood	Charles
Geospiza magnirostris	.	X	X	X	X	X	X	X	X	X	.	.	.
G. fortis	.	.	.	X	X	X	X	X	X	X	X	.	X
G. fuliginosa	.	.	.	X	X	X	X	X	X	X	X	X	X
G. difficilis	A	A	B	B	.	C	C	(C?)	(B?)	.	.	.	.
G. scandens	.	.	.	X	X	X	X	X	X	X	X	.	X
G. conirostris	A	.	B	.	.	.	.	.	.	.	.	C	.
Camarhynchus crassirostris	.	.	.	X	X	X	X	X	X	.	X	.	X
C. psittacula	.	.	.	A	A	B	B	C	C	B	.	.	B
C. pauper	.	.	.	.	.	.	.	.	.	.	.	.	X
C. parvulus	.	.	.	(A?)	.	A	A	A	A	A	B	.	A
C. pallidus	.	.	.	.	.	A	A	A	.	.	B	.	(A?)
C. heliobates	.	.	.	.	.	.	.	X	X	.	.	.	.
Certhidea olivacea	A	A	B	C	C	D	D	D	D	E	F	G	H
Pinaroloxias inornata	Cocos Island												

Notes. (i) 'X' denotes a resident breeding population, and the other letters a division into island forms as follows:

Geospiza difficilis:
A = *septentrionalis* Rothschild and Hartert; B = *difficilis* Sharpe; C = *debilirostris* Ridgway.

Geospiza conirostris:
A = *darwini* Rothschild and Hartert; B = *propinqua* Ridgway; C = *conirostris* Ridgway.

Camarhynchus psittacula:
A = *habeli* Sclater and Salvin, B = *psittacula* Gould; C = *affinis* Ridgway.

Camarhynchus parvulus:
A = *parvulus* (Gould); B = *salvini* Ridgway.

Camarhynchus pallidus:
A = *pallidus* (Sclater and Salvin); B = *striatipectus* (Swarth).

Certhidea olivacea:
A = *becki* Rothschild; B = *mentalis* Ridgway; C = *fusca* Sclater and Salvin; D = *olivacea* Gould; E = *bifasciata* Ridgway; F = *luteola* Ridgway; G = *cinerascens* Ridgway; H = *ridgwayi* Rothschild and Hartert.

(ii) One adult *Geospiza difficilis debilirostris* collected on South Albemarle, two adults probably referable to *G. difficilis difficilis* from Narborough, three *Camarhynchus parvulus* from Abingdon and one *C. pallidus* from Charles are very probably representatives of small breeding populations on these islands, but the numbers collected are too small for this to be considered certain, so they are listed with a query in the table.

(iii) *Geospiza scandens* defies consistent nomenclatural treatment; the forms on the adjacent islands of James and Bindloe are completely separable, so should receive separate subspecific names, but the gap between them is bridged by the populations on islands south of both of them. Hence all the island forms have hesitatingly been united under one name.

(iv) If Darwin's large specimens of *Geospiza magnirostris* are accepted as belonging to an extinct form from Charles (see p. 22), then the specimens of this species from all other islands ought probably to be referred to another subspecies, *G. magnirostris strenua* Gould

STRAGGLERS

So much collecting has been carried out that, apart from the doubtful cases in note (ii), Table IV can probably be taken as a complete list of the finch species breeding on each Galapagos

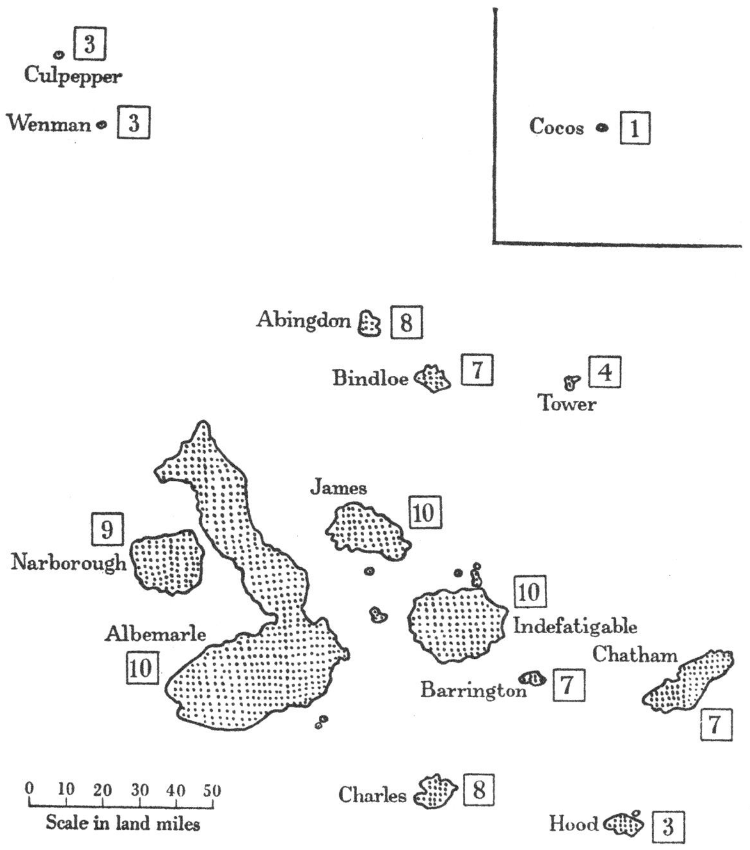

Fig. 4. Number of species of Darwin's finches on each island.

island. In addition there are a few cases, not listed in the table, in which finches have wandered from one island to another without becoming established. The Academy expedition caught a juvenile of the cactus ground-finch *Geospiza scandens* at sea 20 miles south of Indefatigable, and have four records of the medium ground-finch *G. fortis* seen a few miles offshore. The same expedition took fifteen juvenile specimens of *G. fortis* on Hood. No other collectors have found this species on Hood, and we saw none there, despite careful search, so the Academy birds were presumably stragglers. The same probably applies to one specimen of *G. fortis*, four of the small ground-finch *G. fuliginosa*, and one of the small insectivorous tree-finch *Camarhynchus parvulus*, taken by the Academy expedition on Wenman, as all were juveniles, and these species have not otherwise been found there. Wenman is so small that, if these species had been resident, they could scarcely have been overlooked by other collectors. Another presumed straggler in the Academy collection is a typical specimen of the large ground-finch *Geospiza magnirostris* from Charles, where no other recent expedition has found this form. The occurrence of these stragglers, nearly all juvenile individuals, is of importance when considering the degree of isolation of island forms.[1]

EXTINCT FORMS

Nearly all the finches collected by Charles Darwin are similar in appearance to those taken by later collectors, but there are two forms which have not been recorded since 1835. First, there are three male and four female specimens obviously referable to the large ground-finch *Geospiza magnirostris*, but which are considerably larger than any collected since. They are labelled as coming from Charles, but Swarth (1931) thought that this was probably a mistake, and concluded that the form of this species must have changed since the time of Darwin's visit. However, while it would be pleasing to demonstrate measurable evolution on the basis of specimens collected by Darwin, it seems far more probable that these large birds represent an extinct subspecies of *G. magnirostris* from Charles, where the bird no longer resides. Darwin collected typical present-day *G. magnirostris* on James, and expressly noted that the larger form did not occur there. Further, though Darwin unfortunately mingled some of the specimens from different islands, his remarks (1839) make it

[1] There are some further specimens of Darwin's finches, collected by other expeditions, which certainly do not represent breeding populations on the islands where they were found, and so may either be stragglers or are incorrectly labelled. These are one adult *Geospiza fortis* labelled as from Tower, one *G. scandens* of the Bindloe type from James, one *G. c. conirostris* from Gardner near Charles (which suggests an error for Gardner near Hood, where this form occurs), and two *Camarhynchus psittacula* from Chatham.

clear that this applied only to specimens from the first two islands which he visited, namely, Chatham and Charles. When on Charles, Darwin came to realize that specimens from different islands might differ in appearance, and thereafter he kept the collections from each island separate. Hence specimens labelled as coming from either Chatham or Charles were presumably collected on either one or the other. Darwin's specimens most probably came from Charles, since he spent only a few hours ashore on Chatham, but several days on Charles. Further, Charles is a small island and was intensively settled just before Darwin's visit. Probably as an indirect result of this settlement, the endemic mockingbird *Nesomimus* has become extinct on Charles, and the same factor possibly brought about the extinction of the Charles form of *Geospiza magnirostris*.

Also among the *Beagle* specimens are two which in my opinion belong to an unknown form related to the sharp-beaked groundfinch *G. difficilis*. They have a similar shape of beak, though the beak is larger. Gould named one of these specimens *G. nebulosa*, and, I think mistakenly, referred a second rather similar specimen to another species. The one named *G. nebulosa* is labelled as coming from Charles, so it may well represent an extinct form of *G. difficilis* from this island, as the latter species does not occur there at the present time. It may be added that Darwin also collected a typical specimen of *G. difficilis*, presumably on James, as it belongs to the James form of this species.

INTERMEDIATE SPECIMENS

Rumour has it that the gardens of natural history museums are used for surreptitious burial of those intermediate forms between species which might disturb the orderly classifications of the taxonomist. Actually, specimens intermediate between two species are extremely rare in birds, and almost every specimen can be immediately assigned to a known species, except in a few cases where an obvious hybrid is involved. But in this respect Darwin's finches are exceptional, and specimens intermediate between two species, though rare, are less rare than usual. As discussed in Chapter x, some of these intermediates are probably freaks, and others are possibly of hybrid origin. Their existence has added to the difficulties of the systematist in correctly determining the species of Darwin's finches.

CHAPTER III: ECOLOGY

The several species of Geospiza are undistinguishable from each other in habits.... They frequent the rocky and extremely arid parts of the land sparingly covered with almost naked bushes, near the coasts....I seldom, however, saw these birds in the upper and damp region, which supports a thriving vegetation.

CHARLES DARWIN: *The Zoology of the Voyage of H.M.S. 'Beagle'*

HABITAT DISTRIBUTIONS

IN Europe and North America, closely related species of song-birds tend to occupy different habitats, and one of the first questions investigated during our visit to the Galapagos was the extent to which the same might apply in Darwin's finches. The Galapagos provide three main habitats suitable for passerine birds, first an arid lowland zone, secondly a humid forest zone, and thirdly a humid zone of open uplands, while between the arid lowlands and the humid forest is a large area transitional in type. A fourth habitat, coastal mangrove swamp, occupies extensive areas on Albemarle, but only small areas on other islands. The distribution of Darwin's finches in these habitats is summarized in Table V.

GENERIC DIFFERENCES

Table V shows that there is a broad difference in breeding habitat between the ground-finches *Geospiza* and the tree-finches *Camarhynchus*, since the former (omitting *G. difficilis*) breed in the arid zone, but not in the humid zone, while the latter (omitting *C. heliobates*) breed in the humid zone and only rarely in the arid zone. But their separation is very incomplete, since the members of both genera breed side by side in the extensive areas occupied by the transitional zone. Outside the breeding season the habitat difference largely disappears, since the ground-finches then feed regularly in the open uplands and in open parts of the humid forest, while the tree-finches become common in the arid lowlands.

There is also a broad difference in feeding habitat between the two genera, since the species of *Geospiza* tend to feed on the ground and those of *Camarhynchus* in the trees. But all the tree-finches also feed regularly on the ground, while in the breeding season the ground-finches feed commonly in the trees.

The warbler-finch *Certhidea* has a wider breeding range than the other genera, since it breeds commonly throughout the arid, transitional and humid forest zones, and also on the uplands where there are bushes or thick bracken. It is also the only one of Darwin's finches to occur on all the main islands, and nearly everywhere it is the most numerous species.

TABLE V. BREEDING AND FEEDING HABITATS OF DARWIN'S FINCHES

Species	Chiefly breeds in	Chiefly feeds on or in
GEOSPIZA		
magnirostris	Arid and transitional	Ground
fortis	Arid and transitional	Ground
fuliginosa	Arid and transitional	Ground
difficilis	Varies with island form	Ground (and *Opuntia*)
scandens	Arid and transitional with *Opuntia*	*Opuntia*
conirostris	Arid	Ground and *Opuntia*
CAMARHYNCHUS		
crassirostris	Transitional and humid	Trees
psittacula	Transitional and humid	Trees
pauper	Transitional and humid	Trees
parvulus	Transitional and humid	Trees
pallidus	Transitional and humid	Trees
heliobates	Coastal mangroves	Mangroves
CERTHIDEA		
olivacea	Arid, transitional and humid	Bushes and trees
PINAROLOXIAS		
inornata	Cocos forest	Trees and ground

Notes. (i) The variations in *Geospiza difficilis* are set out in Table VI, p. 27.
(ii) While all the species of *Camarhynchus* (except *C. heliobates*) breed mainly in the transitional and humid zones, they are occasionally found singing in the arid lowlands. *C. crassirostris* and *C. parvulus* were proved to nest there in small numbers, and the other species perhaps do so locally or in small numbers. In the humid zone these species extend to the tree limit, but except for *C. pallidus* they are scarce above the forest.

None of Darwin's finches has become adapted for breeding on the open uplands away from thick cover. Although this group has produced such unfinchlike forms as the warbler-finch and the woodpecker-finch, there is not, as yet, a 'lark-finch'. This is somewhat curious, as in other parts of the world finches have become adapted to open country away from all bushes and trees. The latter, for instance, is true of several European and North American buntings and longspurs. Since the Galapagos uplands occupy only a very limited area, there is perhaps scarcely room for a special form to evolve there.

CLOSELY RELATED SPECIES

The closely related species of Darwin's finches do not in most cases occupy separate habitats. Thus the four common species of *Camarhynchus* breed side by side in the transitional and humid forest. This applies to the vegetarian tree-finch *C. crassirostris*, the two insectivorous tree-finches *C. psittacula* and *C. parvulus*, and the woodpecker-finch *C. pallidus*. In only one case are two species separated in habitat, the mangrove-finch *C. heliobates* breeding exclusively in the coastal mangrove belt and the closely related woodpecker-finch *C. pallidus* inland; the former species is confined to Albemarle and Narborough.

Similar considerations hold in the ground-finches *Geospiza*. The habitats of the large *G. magnirostris* and the medium *G. fortis* seem identical, while that of the small *G. fuliginosa* is extremely similar. As compared with the two former species, *G. fuliginosa* is slightly less tolerant of dense transitional forest and slightly more tolerant of low trees, but this makes little difference in practice, and the three species nearly everywhere breed side by side. The fourth species, the cactus ground-finch *G. scandens*, has a more restricted habitat, since it is confined for breeding to the prickly pear *Opuntia*. But prickly pear trees occur commonly throughout the arid and transitional zones on most islands, so that *Geospiza scandens* is correspondingly widespread, and it is not segregated from the other three ground-finches, since all of these nest freely in *Opuntia*.

The fifth species of *Geospiza*, the sharp-beaked ground-finch *G. difficilis*, is peculiar since it differs in habitat on different islands, as set out in Fig. 5 and Table VI, opposite.

Table VI shows that on the central islands of James, Indefatigable and Abingdon, the sharp-beaked ground-finch *G. difficilis* breeds only in the humid forest, whereas on the outlying islands of Culpepper, Wenman and Tower it breeds in the arid lowland zone, which is the only zone present. As discussed in Chapter XI, *G. difficilis* is probably an older form than the small ground-finch *G. fuliginosa*, and its peculiar habitat distribution is best explained by supposing that it has been eliminated by *G. fuliginosa* wherever the two species have come in contact. *G. fuliginosa* breeds on all the main Galapagos islands except the outlying Culpepper, Wenman and Tower, and only on these latter

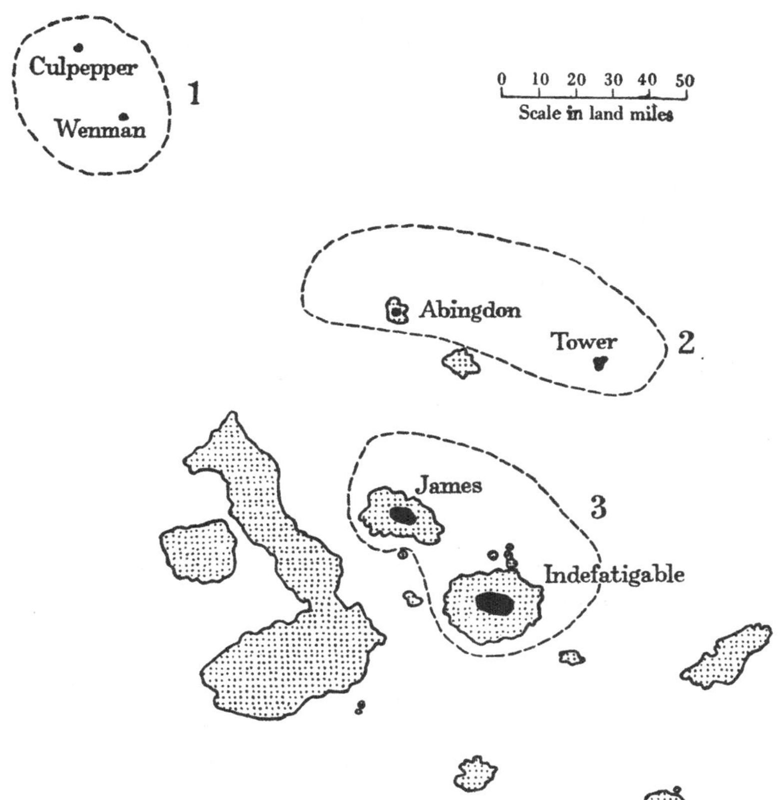

Fig. 5. Distribution of *Geospiza difficilis* (black) in relation to that of *G. fuliginosa* (stippled).
The interrupted lines show the distribution of the three races of *G. difficilis*: 1, *septentrionalis*; 2, *difficilis*; 3, *debilirostris*.
For details, see Table VI below.

TABLE VI. BREEDING ZONE OF *GEOSPIZA DIFFICILIS*

Subspecies	Island	Zone	Comment
G. d. septentrionalis	Culpepper and Wenman	Arid	No other zones present. *G. fuliginosa* absent
G. d. difficilis	Tower		
G. d. difficilis	Abingdon	Humid	*G. fuliginosa* breeds in arid and transitional zones
G. d. debilirostris	James and Indefatigable		

Notes. (i) The Culpepper and Wenman form differs from the others in feeding on *Opuntia*.
(ii) *G. difficilis* may also breed in small numbers in the humid forest of South Albemarle and on the high ground on Narborough (see p. 20).

islands does *G. difficilis* breed in the arid zone. However, *G. fuliginosa* seems unable to breed in humid forest, and *G. difficilis* also breeds on those Galapagos islands which have humid forest, namely, James, Indefatigable and Abingdon (and possibly Albemarle and Narborough, see p. 20).

The distribution of *G. difficilis* in the Galapagos presents a close parallel with that of the bank vole *Evotomys* (now *Clethrionomys*) *nageri* in western Europe, discussed by Barrett-Hamilton and Hinton (1913–14). In western Europe *C. nageri* occurs primarily at higher altitudes, being replaced at lower altitudes by a related species *C. glareolus*. But geographical forms of *C. nageri* live at low levels on a few outlying European islands, such as Jersey, Skomer, Mull and Raasay, where *C. glareolus* is unknown. The evidence suggests that in late pleistocene times *C. nageri* was widespread at lower altitudes, and that its more restricted distribution at the present day is due to competition with the later arrival *C. glareolus*. At the latitude of Britain and France, *C. nageri* survives at low altitudes only on those islands which *C. glareolus* has not reached.

LIMITS TO HABITAT DISTRIBUTION

The preceding survey shows that, except in two cases, the closely related species of Darwin's finches are not separated from each other in habitat. They are thus in marked contrast to many European passerine birds. For example, three species of pipits breed in Britain, the rock pipit *Anthus spinoletta* on rocky shores, the meadow pipit *A. pratensis* in open country inland, and the tree pipit *A. trivialis* in open country with trees. In the main the three species breed apart from each other, and they come in contact only where their respective habitats adjoin. A general survey of British breeding song-birds provides many similar examples (Lack, 1944*a*).

It has sometimes been assumed that in such cases the chief factor which prevents one species from breeding in the habitat of a closely related form is a difference in food requirements. However, the species concerned often eat very similar foods. For example, in Britain the reed bunting *Emberiza schoeniclus* breeds primarily in marshes and eats a wide variety of marsh plants and insects, while the yellow bunting *E. citrinella* breeds in dry scrub and eats a similar variety of plants and insects of drier country. All the evidence indicates that this small difference in diet is the result, and not the cause, of the habitat difference, particularly bearing in mind the varied nature of the foods taken by each species. A parallel argument applies in many other cases.

Habitat differences have also been attributed to climatic factors, temperature, humidity and the like. But birds are warm-blooded and comparatively independent of their physical environment. Climatic differences could scarcely account for the restriction of the reed bunting to marshes and the yellow bunting to drier country in the same region, or of the meadow pipit to open treeless country and the tree pipit to open country with trees in the same region. Obviously a species can breed only where food and climate permit, and certain habitats are thereby excluded, but most birds have a specific habitat which is far more restricted than can be accounted for solely through the direct effects of food and climate.

Field study has shown that each species of bird selects its habitat, recognizing it by features which are not necessarily those directly essential to its existence (Lack, 1933). Such habitat selection has survival value, since if a bird breeds in the habitat of its ancestors it will probably be successful, while if it tries elsewhere the habitat may be unsuitable. At one time I supposed that the restricted habitat of each species was due primarily to the rigidity of its behaviour, and that, could it break away from the latter, it might be successful in other habitats. But further consideration has led to my abandoning this view. Just as a bird is occasionally coloured abnormally, or lays freak eggs, so it occasionally breeds outside its customary habitat. But these irregularities do not normally persist, which strongly suggests that they are eliminated by natural selection.

Rarely, a species is established locally in an unusual habitat, and such instances throw valuable light on the general problem. One case occurs in Darwin's finches, since on three outlying islands the sharp-beaked ground-finch *Geospiza difficilis* breeds in the arid lowlands, but it does not frequent this habitat on the central islands. As already discussed, this seems due to competition with the related *G. fuliginosa*. In general, competition between species seems one of the most important factors limiting habitat distribution in birds. A particularly good example is cited by Streseman (1939). In Burma, where there is no species closely related to it, the white-eye *Zosterops palpebrosa* breeds from sea-level to the high mountains. But in Malaya and Borneo it is restricted to the lowlands, and a related species of *Zosterops* occupies the middle and higher zones. In Java, Bali and Flores, one closely related species of *Zosterops* occupies the higher regions, and another the coast, and here *Z. palpebrosa* is restricted to the middle altitudes. These differences in habitat in different regions are obviously correlated with the presence or absence of related species.

Another case of a local change in habitat occurs in the Galapagos, where the warbler *Dendroica petechia* breeds from the coast inland to the humid forest, whereas in Ecuador and Colombia it is confined to the coastal mangroves. The latter restriction is perhaps due to the presence inland of warblers or other birds which are absent from the Galapagos, but this view must be regarded as conjectural until further evidence has been obtained. It is interesting that *D. petechia* also breeds inland on some of the small West Indian islands, but on Jamaica and other large islands it is, as in Ecuador, confined to the coastal mangroves.

Other cases of habitat variation on islands are summarized in a general survey published elsewhere (Lack, 1944*a*). In the same paper instances are given in which two mainland races of the same species differ in habitat, such as the shore and alpine forms of the pipit *Anthus spinoletta*, the marsh and dry country forms of the song sparrow *Melospiza melodia*, and the woodland and moorland forms of the ring ousel *Turdus torquatus*. In most such cases, the reason for the differences concerned is not known. That, as these instances show, birds can at times modify their choice of habitat, is important from the evolutionary standpoint, since environments do not stay permanently the same.

To return to Darwin's finches, in these birds, unlike continental land birds, the closely related species do not in most cases occupy separate habitats. This difference is connected with other ecological factors to be discussed in Chapter VI, and further consideration is postponed until then. The significance of habitat differences in regard to the origin of species is another important problem which cannot be discussed until later in the book.

NESTS

Darwin's finches build bulky nests, cup-shaped below, with a domed roof and a side entrance. The roof is perhaps of value in shading the eggs from the sun. Such roofed nests are common in tropical weavers and in Central American finches, but are rare in such birds outside the hot regions, so that when this habit is present, as in the house sparrow *Passer domesticus*, it suggests that the species may be of tropical origin.

The nests of all the species of Darwin's finches are extremely similar, including that of the warbler-finch *Certhidea*. Those of the larger species tend to be slightly bulkier than those of the smaller species, but there is an extremely wide overlap in this respect. The nest materials are also similar in all the species. In the transitional and humid zones, twigs, grasses and large quantities of an epiphytic grey lichen are used, while in the arid lowlands, where the lichen does not grow, the nests are made chiefly of twigs, grasses and wild cotton.

The nests are placed from 3 to 30 ft. above the ground, and there is little if any significant difference in the height selected by each species. Indeed, there is only one important respect in which different species differ in regard to nesting, this being in the extent to which the prickly pear *Opuntia* is used. The cactus ground-finch *Geospiza scandens* nests exclusively between the terminal pads of a prickly pear tree. The other ground-finches commonly use the same situation, but in addition they frequently nest among small, closely growing twigs of *Acacia*, *Maytenus* and many other bushes and trees, usually at the end of a branch. The tree-finches *Camarhynchus* and the warbler-finch *Certhidea* nest in the latter type of situation, but were never found nesting in prickly pear. The characteristic ground-finch nest in *Opuntia* is shown in Plate II.

In all of Darwin's finches the nest features prominently in display, both between an unmated cock and a passing hen and between a mated male and his mate. Although monogamous, the cock not only builds several nests, but also builds and displays regularly at nests which have been started by, or are still being used by, individuals of other species, either for display or breeding. For example, a male warbler-finch *Certhidea* frequently visited, displayed and built at a nest largely constructed by a medium ground-finch *Geospiza fortis*, which was also using it for display. Elsewhere, the reverse was witnessed. Again, during two days' observations of one courting male of the small ground-finch *G. fuliginosa*, this bird visited, built and displayed regularly at eight nests, some of which were also visited by males or pairs of four other species, namely, the medium ground-finch *G. fortis*, the tree-finches *Camarhynchus crassirostris* and *C. parvulus*, and the warbler-finch *Certhidea*. Again, at the unexpected arrival of a female of its species, an unmated male *Camarhynchus parvulus* immediately displayed at and entered the nearest available nest, which happened to be one in which a hen woodpecker-finch *C. pallidus* was incubating; the latter promptly chased the intruder out. Such incidents were repeatedly observed. Not only are the nests of other species used regularly for display, but for their final nest a pair often take over a nest built largely by members of another species. While display-building is carried out primarily by the cock, the final nest is usually built partly, if not mainly, by the hen.

Nesting sites clearly provide no means of isolation between the various species of Darwin's finches. Indeed, there are no other birds in which the nests of other species are taken over so frequently and so cavalierly, since often they are just 'borrowed' for a particular occasion. Such a situation is possible because all the species have similar nesting requirements, and it is connected with the important part played by the nest in courtship.

EGGS

The eggs of the various species differ somewhat in size, but otherwise look much alike, being white with small pinkish spots. The usual clutch is four, and almost all the broods seen out of the nest consisted of four young. We found no evidence for extensive sterility among the eggs, which was claimed by Beebe (1924). One pair of the small ground-finch *Geospiza fuliginosa* raised five young.

It is sometimes claimed that the number of eggs which a bird lays is regulated primarily by the mortality of the species, and that the upper limit of clutch size is nicely adjusted so as to avoid over-population. But it is extremely difficult to see how such an adjustment can be brought about through natural selection, since the latter operates primarily for the survival of the individual, or of the individual family, and only incidentally for the good of the species.

It seems much more likely that in birds such as Darwin's finches, which feed their young in the nest, clutch size is fixed by the maximum number of young which the parents can successfully raise at one time. Parents which raise fewer than the maximum number leave fewer offspring than those which raise the maximum, while when the maximum is exceeded, the brood tend to starve before leaving the nest, so that there are no survivors. This supposition is corroborated by the fact that passerine and other birds tend to lay larger clutches and to raise larger broods the longer the hours of daylight, and so of available feeding time, in their breeding quarters. For example, Boyd (1936) has shown that the swallow *Hirundo rustica* lays on the average a larger clutch in the north than in the south of Britain.

BREEDING SEASON

All the Galapagos finches breed during the rainy season, which normally occurs between about December and March. In the transitional zone, where the rain starts about a month earlier than on the coast, breeding also starts about a month earlier. The birds raise a succession of broods until the end of the rainy season and then stop. If the rains continue abnormally long, breeding continues likewise. Thus, in the exceptionally wet year of 1932, Swarth (MS. diary) found that breeding continued into June. At least in some years a little breeding takes place outside the rainy season, perhaps only when rain happens to fall in this period. For instance, Gifford (1919) recorded some breeding on Abingdon during a rainy spell in September. In our captive ground-finches brought back to California, Dr R. T. Orr found that song was stimulated by the falling of rain, as it is in the wild.

No species of Darwin's finches are isolated from each other by breeding season. The cactus ground-finch *Geospiza scandens* and the woodpecker-finch *Camarhynchus pallidus* start to breed a few weeks earlier than the other species in their respective habitats, but both species continue to breed for several further weeks alongside the others. The other species start to breed at about the same time as each other.

The adult finches normally moult after the breeding season, in May and June. To judge from collected specimens, the moult occasionally takes place at other times, perhaps when breeding has occurred outside the normal season.

ENEMIES

Darwin's finches have extremely few natural enemies. Gifford (1919) and Beebe (1924) record the Galapagos short-eared owl *Asio galapagoensis* as eating both the medium and the small ground-finch, *Geospiza fortis* and *G. fuliginosa.* This owl also subsists on rice-rats, lizards, grasshoppers and crabs, and it is doubtful whether ground-finches form a big part of its diet. The Galapagos barn owl *Tyto punctatissima* has not been recorded taking any of the finches, so is probably not a serious enemy, though it may be suspected that it takes an occasional individual. The European representatives of these two owls feed primarily on rodents, but to a minor extent on small birds. The only Galapagos hawk, *Buteo galapagoensis,* is almost certainly harmless to the adult finches, though it might pick up an occasional fledgling. Nor have any of the native Galapagos mammals or reptiles been recorded eating the finches. The larger reptiles, it will be recalled, are vegetarian.

While the adult finches appear to have no serious natural enemies, the fact that their nests are placed inaccessibly at the ends of branches suggests that they may have potential nest enemies. Perhaps the rice-rats, the snakes or the small lizards might take their eggs if they could reach them. But at the ends of branches the nests seem safe enough. The latter conclusion is supported by the fact that the nest features prominently in display, a habit normally found only in birds, such as herons or cormorants, which have no nest enemies.

As compared with the birds of civilized countries, Darwin's finches are remarkably tame. The naturalist can usually walk to within 10 ft. of them, though, like other birds, they are somewhat warier near their nests. We also observed that when several of the vegetarian tree-finch *Camarhynchus crassirostris* were feeding on fallen fruit, a tame cat caught one of their number and scarcely disturbed the rest. Some other Galapagos birds are tamer than the finches. Thus the hawk not infrequently allows itself to be touched, the mockingbird *Nesomimus* sometimes pecks at the eyeholes of one's boots, and individuals of the tyrant flycatcher *Myiarchus magnirostris* tried to remove the hair from our heads and bodies as nesting material. In former times the birds were even tamer, and William Dampier records the dove *Nesopelia galapagoensis* settling in clusters on the hats and shoulders of his fellow buccaneers. Such tameness is general among birds which do not know man or other large enemies, but is gradually lost on those desert islands later colonized by man, a fact pointed out for the birds of the Falkland Islands by Darwin (1839). It is a curious pleasure to have the birds of the wilderness settling upon one's shoulders, and this pleasure could be much less rare were man less destructive.

Dr R. T. Orr found that, compared with Californian song-birds, our captive ground-finches were weak in flight and clumsy in manœuvre, presumably because, in the Galapagos, selection is not maintained by fast-flying birds of prey. Yet, surprisingly, the captive finches showed marked fear of Californian hawks. Also, we on one occasion saw some ground-finches, chiefly the small *Geospiza fuliginosa*, mobbing a Galapagos short-eared owl. Perhaps, after all, the owl is a serious enemy, but more probably the finches have retained a response to predators which has little or no relevance to life in the Galapagos. Similarly, when Swarth (1935) put a Galapagos dove *Nesopelia* off its nest, it 'feigned injury', although the bird apparently has no natural enemies.

In Darwin's finches each pair normally raises at least one brood of four young per year, and as the birds do not increase indefinitely, there must be a heavy mortality, but it is not at present known how the majority of the birds die. The only published record of heavy mortality is a note by Captain Colnett that in June 1793 he found great numbers of birds dead in their nests, but this appears to relate to fledglings rather than to adult birds. The only record of disease is a note by Gifford (1919) that individuals of *Geospiza fortis*, *Camarhynchus crassirostris* and *Certhidea* were suffering from diseased feet. The most likely causes of adult mortality among the finches are periodic food shortage or disease, in which case their numbers are perhaps subject to marked fluctuations, but this interesting problem can be studied only by long residence in the islands.

CHAPTER IV: FEMALE PLUMAGE

When both sexes are obscurely coloured...and when no direct evidence can be advanced showing that such colours serve as a protection, it is best to own complete ignorance of the cause.

CHARLES DARWIN: *Selection in Relation to Sex*, Ch. XVI

DIFFERENCES BETWEEN SPECIES

IN all of Darwin's finches the females are dully coloured, the plumage differences between the species are in most cases slight, while some of the closely related species have identical plumage. The latter holds, for instance, in the case of the large, medium and small ground-finches *Geospiza magnirostris*, *G. fortis* and *G. fuliginosa*, suggesting that these species may have diverged from each other only in comparatively recent times. In these three species, the upper parts are greyish brown with dark streaks, and the under parts are much paler with fairly heavy dark streaks. Their plumage forms a convenient standard with which to compare the other species, as set out in Table VII.

DARK STREAKING

In Darwin's finches dark streaking is probably primitive, as it occurs in the more finch-like genus *Geospiza*, but is reduced or absent in the tree-finches *Camarhynchus* and in the warbler-finch *Certhidea*, which have diverged more from the finch-like type. Further, though adults of the warbler-finch *Certhidea* are unstreaked, juveniles of the Charles form are heavily streaked; as is well known, juvenile animals frequently show primitive characters which have been lost in the adults. In all the other forms of Darwin's finches, so far as known, the juveniles have similar coloration to the adult females.

It is not certain what advantage, if any, is associated with the loss of dark streaking in the tree-feeding genera *Camarhynchus* and *Certhidea*. These birds are also paler on the upper parts than are the female ground-finches, and as the ground provides a darker background than the trees, protective coloration is possibly involved.

TABLE VII. SPECIFIC DIFFERENCES IN FEMALE PLUMAGE

Species	Upper parts	Under parts	Rufous wing-bar
GEOSPIZA			
magnirostris	Standard	Standard	—
fortis	Standard	Standard	—
fuliginosa	Standard	Standard	—
difficilis	Darker	More streaked	Present in two races, rare in third
scandens	Dark as *difficilis* and greyer	As *difficilis*	—
conirostris	Yet darker	Yet more streaked	Sometimes in one race
CAMARHYNCHUS			
crassirostris	More olive-brown, less streaked	More buff	—
psittacula	Paler, more olive-grey, much less streaked	Much less streaked	—
pauper	Darker and browner than *psittacula*	More streaked than *psittacula*	—
parvulus	As *psittacula*	As *psittacula*	—
pallidus	Yet paler, almost unstreaked	Almost or quite unstreaked	—
heliobates	Dark as standard, rather less grey, much streaked	Somewhat less streaked than standard	—
CERTHIDEA			
olivacea	Paler, colour variable, unstreaked	Unstreaked	Present in some subspecies
PINAROLOXIAS			
inornata	Somewhat darker than *difficilis* and more olive	Somewhat paler than *difficilis*, more buff, and somewhat less streaked	Present

Note. The above descriptions also apply to the males, except for those which have partly or wholly black feathering (discussed in Chapter v).

Another possibility worth testing is Gloger's rule, which states that the birds of more humid environments have darker plumage than those of drier climates. The reason for this correlation is by no means clear, but it is fairly common in birds. Examples are the darker coloration of forest forms as compared with those of open country in West Africa (Bates, 1931), and the darker plumage of certain British song-birds as compared with the continental races of the same species. A similar correlation does not hold among Darwin's finches. Most species of tree-finch *Camarhynchus* breed in the humid forest and most species of ground-finch *Geospiza* in the arid lowlands, but the former have paler, not darker, plumage than the latter. Only one species of *Camarhynchus*, the mangrove-finch *C. heliobates*, breeds in the lowlands, and this bird has darker, not paler, plumage than the other species. The ground-finch with the darkest female plumage is the large cactus ground-finch *Geospiza c. conirostris*, which is confined to the particularly arid lowlands of Hood. Another dark species is the sharp-beaked ground-finch *G. difficilis*, of which one race breeds in the humid forest of James and Indefatigable and another in the arid lowlands of Culpepper and Wenman, but the former is no darker than the latter. Clearly, Gloger's rule does not apply in Darwin's finches.

INDIVIDUAL VARIATION

While the degree of streaking is characteristic for each species and race of Darwin's finches, there is also a large amount of purely individual variation. As a result, the specific and racial characters are true only as general tendencies. For example, females of the medium and small ground-finches *Geospiza fortis* and *G. fuliginosa* are usually much streaked, but a few individuals are as little streaked as some forms of *Camarhynchus*, while a few others are streaked as heavily as females of the sharp-beaked ground-finch *Geospiza difficilis* and the cactus ground-finch *G. scandens*. The amount of streaking in the females of the two latter species also varies individually, though usually it is heavier than in the other ground-finches. Again, many of the tree-finches include both some unstreaked and some fairly heavily streaked individuals.

Four species of Darwin's finches show a very small rufous wing-bar, and this also is subject to marked individual variation. Thus, in the Tower and Abingdon race of the sharp-beaked ground-finch *G. difficilis*, only one collected individual had a bright rufous bar, in a few specimens the bar was dull rufous, in most it was brown, and in a few buff. In the other races of this species, the wing-bar is equally variable, though more often rufous. Likewise in the warbler-finch *Certhidea*, both the colour and size of the bar vary markedly, and the proportion of individuals with a rufous bar is different in each island form. So far as known, this rufous wing-bar is functionless, and its sporadic occurrence in four widely different species (see Table VII) suggests that it is a primitive character in process of being lost. In the other species, the edges of the wing coverts are buff or dull brown.

Most passerine birds seem less variable in plumage than Darwin's finches, implying that selection is more rigid. This is possibly because most song-birds have a larger number of natural enemies than Darwin's finches. However, the latter reason is by no means certain. Most of the birds which have no animals preying on them, such as the birds of prey and the larger species of water birds, also show less individual variation in plumage than do Darwin's finches.

DIFFERENCES BETWEEN ISLAND FORMS

In Darwin's finches, small differences in colour of female plumage occur not only between some of the species, but also between island forms of the same species. These are most marked in the warbler-finch *Certhidea*, as set out in Table VIII and Fig. 6, are less marked in some of the other species, as shown in Table IX and Fig. 7, and are not found in the vegetarian tree-finch *Camarhynchus crassirostris* or in the three ground-finches *Geospiza magnirostris*, *G. fortis* and *G. fuliginosa*.

In some birds the differences in colour between races of the same species are adapted to differences in the environmental background. This has been shown most convincingly for various larks, such as *Galerida*, *Mirafra*, *Spizocorys* and *Otocoris*, in which a change in the colour of the background nearly always involves a parallel change in the colour of the larks, while observers have tried in vain to chase the birds off the type of ground to which their plumage is adapted (Niethammer, 1940; Behle, 1942).

As summarized in Fig. 6, the warbler-finch *Certhidea* shows differences in colour on different Galapagos islands, being browner on one island, greener on a second, greyer on a third and very pale on a fourth. That these differences may be adaptive is suggested by the fact that the most olive forms of *Certhidea* are those on Chatham, Indefatigable, James and Albemarle. These islands possess much more forest, and so a greener environment, than the more barren outlying islands. On islands of the latter type *Certhidea* is usually greyer or browner in colour. On Barrington, *Certhidea* is unusually pale, and this also may be

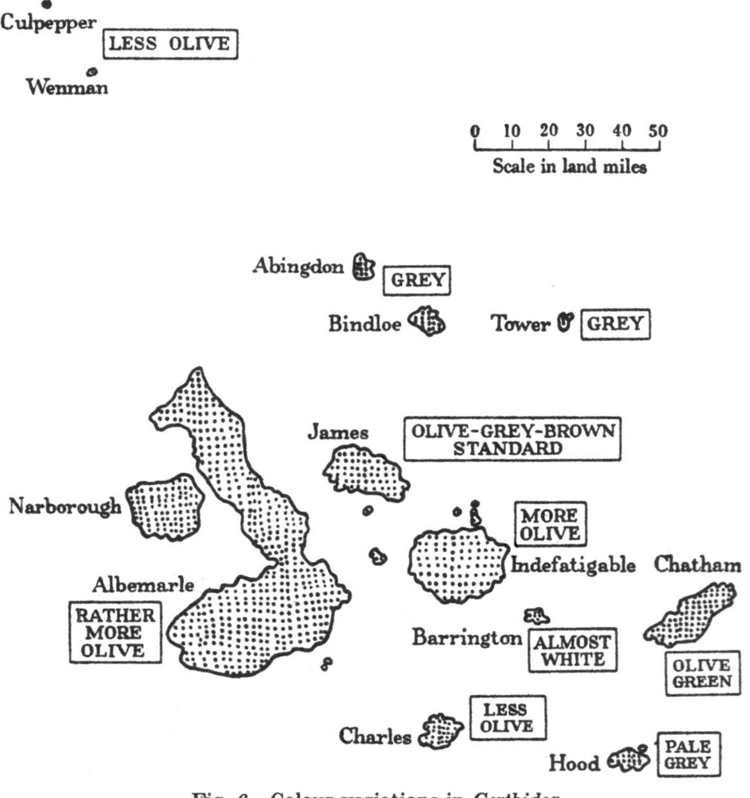

Fig. 6. Colour variations in *Certhidea*.

The birds on James have been taken as standard. For details, see Table VIII opposite.

due to protective coloration, as Gifford (1919) reported that the birds were most difficult to see there. Further, several other Galapagos animals have unusually pale forms on Barrington, such as the mockingbird *Nesomimus* (Swarth, 1931) and the two grasshoppers *Liparoscelis* and *Schistocerca melanocera* (McNeill, 1901; Snodgrass, 1902*b*). However, there are other

TABLE VIII. PLUMAGE VARIATIONS IN *CERTHIDEA OLIVACEA*

Note. The form on James is taken as standard.

Island form	Upper parts	Under parts	Rufous wing-bar
James	Grey-brown with olive tinge	Pale olive-buff	Moderately well developed
Culpepper, Wenman	More grey-brown, less olive	Wash-brown, slightly darker than on Charles	Rufous tint in a few only
Tower	Grey, slightly browner than next	Grey	Absent; rarely brown
Abingdon, Bindloe	Grey	Grey, slightly buff	Absent; rarely brown
Indefatigable	Rather more olive than on James	Rather more olive than on James	Rather better than on James
Albemarle	Slightly more olive than on James	As on James	Maximum development, good in more than half the specimens
Barrington	Very pale grey	White	Absent; rarely a trace of buff
Chatham	As on Indefatigable	More yellow-green than on James	Present in one-fifth of cases, rather fewer than on James
Hood	Pale grey	Slightly paler grey than on Abingdon	Absent; rarely buff-brown
Charles	As on Culpepper	Wash-brown, slightly paler than on Culpepper	Good in a few, but in many buff-brown

Notes. (i) This table holds for males as well as females, except that the males of some forms have an orange-tawny throat patch (see Chapter v).

(ii) The form of *Certhidea* on the most northerly islands, Culpepper and Wenman, comes closest in colour to the form on Charles, the most southerly island, while the form on Abingdon and Bindloe in the north is most similar to that on Hood in the south-east. Hence colour changes have possibly originated on the central islands and spread outwards. An exception to the general regularity is that the Chatham and Indefatigable forms look much alike, but Barrington, in between, has a strikingly different form.

TABLE IX. VARIATIONS IN FEMALE PLUMAGE IN OTHER SPECIES

Islands	Upper parts	Under parts	Rufous wing-bar
	1. *Geospiza difficilis*		
Tower, Abingdon	Less dark	Less streaked	Rarely present
James, Indefatigable	Darker	More streaked	Present
Culpepper, Wenman	Darker and more olive	More streaked and more buff	Present
	2. *Geospiza scandens*		
All except Abingdon and Bindloe	As in Table VII	As in Table VII	—
Abingdon	Darker than most	Darker than on Bindloe	—
Bindloe	Darker still	Darker than most	—
	3. *Geospiza conirostris*		
Tower	Like *G. scandens*	Like *G. scandens*	—
Culpepper	More olive	More buff	Often present
Hood	Much darker	Much darker and more streaked	—
	4. *Camarhynchus psittacula*		
James, Indefatigable, Barrington, Charles	As in Table VII	As in Table VII	—
Abingdon, Bindloe	Slightly darker and greyer, less olive	As in Table VII	—
Albemarle	Slightly darker and browner	Rather more streaked	—
	5. *Camarhynchus parvulus*		
All except Chatham	As in Table VII	As in Table VII	—
Chatham	Browner, more barred	Yellow-green tinge, somewhat more streaked	—
	6. *Camarhynchus pallidus*		
James, North Albemarle	Very grey	Pale grey, normally unstreaked	—
Indefatigable	Very olive	Buff, normally unstreaked	—
South Albemarle	Rather less olive than on Indefatigable	Intermediate between James and Indefatigable	—
Chatham	Somewhat olive, dark bars on head	Streaked	—

colour variations in *Certhidea* which seem without adaptive significance. This applies particularly to the small rufous wing-bar, the extent of which varies markedly, and apparently quite pointlessly, from island to island.

As shown in Table IX, some of the tree-finches *Camarhynchus* also differ in colour on different islands. These variations are extremely difficult to relate to possible differences in environment. The tree-finches live in the humid and transitional forest, which looks very similar on different islands. Further, though

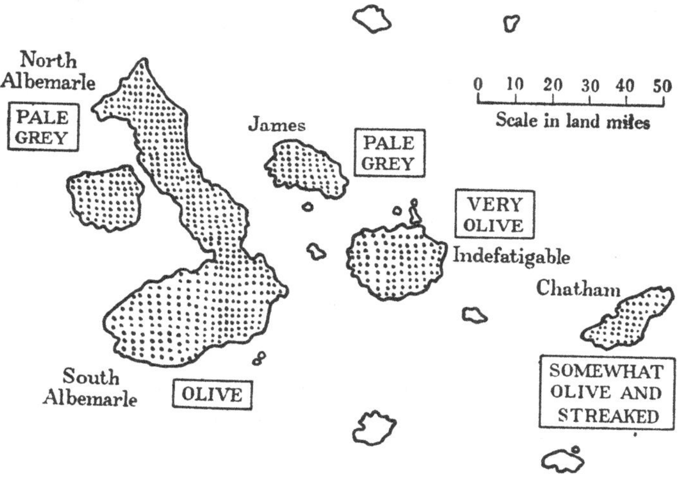

Fig. 7. Colour variations in *Camarhynchus pallidus*.

the various species live side by side in the same habitat, their colour variations do not run parallel with each other. For instance, on Chatham the small insectivorous tree-finch *C. parvulus* is greener than on other islands, but the woodpecker-finch *C. pallidus* is less olive than the rule, while the vegetarian tree-finch *C. crassirostris* is normal. Again, the woodpecker-finch *C. pallidus* is an unusually pale grey on James and North Albemarle, but no other tree-finches are greyer than usual there. Similarly, the large insectivorous tree-finch *C. psittacula* is browner and more streaked on Albemarle, but this does not apply to the Albemarle forms of the other tree-finches. Hence the colour differences in question may be without adaptive significance, but such a negative view is extremely difficult to establish with certainty. Some of the colour differences between island forms of the ground-finches also seem pointless. Thus the proportion of individuals with rufous on the wing-bar is markedly different in the three races of the sharp-beaked ground-finch *Geospiza difficilis*.

To sum up, many tropical birds are brightly coloured. That Darwin's finches are dull and inconspicuous suggests that, though they have few enemies, these have been sufficient to promote some degree of concealing coloration. This is supported by the fact that the ground-finches live against a darker background than the tree-finches and are likewise of a darker colour, while some of the colour differences between island forms of *Certhidea* also seem adaptive. But other colour variations are, so far as known, without adaptive significance, notably the differences between island forms of the tree-finches, and the extent of the rufous wing-bar in *Geospiza difficilis* and *Certhidea*. The possible ways in which non-adaptive differences might arise between island forms are considered in a later chapter.

CHAPTER V: MALE PLUMAGE AND SEXUAL SELECTION

In each species [of *Geospiza*]...a perfect gradation in colouring might, I think, be formed from one jet black to another pale brown. My observations showed that the former were invariably the males....The jet black birds...were in singularly few proportional numbers to the brown ones: I can only account for this by the supposition that the intense black colour is attained only by three-year-old birds.

CHARLES DARWIN: *The Zoology of the Voyage of H.M.S. 'Beagle'*

DISPLAY

IN Darwin's finches, breeding behaviour follows the same general pattern as in the territorial song-birds of temperate regions. Each cock holds a territory in which it sings and from which it drives out other males, while the hen seeks out the cock in its territory to form a pair. Courtship includes five main activities: the cock visits nests and carries building material in the presence of the hen, he feeds the hen by regurgitation, there are aerial chases, also postures on the ground and in the branches, and finally coition. The birds display with excitement and vigour, but the postures are simple and unspecialized, the commonest attitudes being with wings partly opened and rapidly vibrated, or alternatively with wings fully extended and held motionless or slowly waved. Darwin's finches are among the few birds which use very similar postures in display between rival males, in courtship between male and female, and when the fledgling solicits its parents for food.

The significance of these courtship activities is the same as in other song-birds, so comes outside the scope of this book. However, one point is relevant, that the pattern of courtship behaviour and the postures adopted are the same in all the species of Darwin's finches. Nesting habits, incubation and care of the young are also similar throughout the group. Hence, in Darwin's finches there is much greater uniformity in courtship habits than there is, for instance, in the shape of the beak. This illustrates the fact that, in birds, breeding habits and display can be as valuable an indication of affinities as are structural characters.

SONG

The song of the male Darwin's finch has the same functions as in other song-birds, to advertise possession of territory to rival males and to hens in search of mates. In all the species the song is simple, unmelodious and without complex phrases. The song of the larger species somewhat resembles that of the American redwinged blackbird *Agelaius phoeniceus*, but is usually much less musical; there is no European counterpart. The song of the warbler-finch *Certhidea* is somewhat reminiscent of that of a wren *Troglodytes*, but is much feebler and less well phrased.

In Darwin's finches the variations in song between individuals of the same species are marked, while the characteristics of the different species are ill defined. After much practice we could place many individuals in the correct species by their songs, but in many other cases this was extremely difficult or impossible. There was considerable overlap in the character of the song not only between different species, but even between individuals of different genera. Possibly, though there was confusion to the human ear, the birds could always recognize members of their own kind, but in most other parts of the world the naturalist does not experience such difficulty in distinguishing the songs of different species.

Mockingbirds are famous for the beauty and complexity of their singing, but the Galapagos mockingbird *Nesomimus* has a simple, dull and quiet song. Indeed, there is only one good songster among Galapagos birds. This is the warbler *Dendroica petechia*, which looks so similar to the Ecuadorean form of the same species that it is almost certainly a recent colonist of the islands. The characteristic music of the Galapagos forest is not the song of birds but the braying of donkeys. A similar poverty of bird song has been noted on other oceanic islands, so is perhaps a general tendency in such places. If so, the explanation may be similar to that advanced later for the loss of bright male plumage, but further observations are needed to establish this conclusion in the case of song.

BLACK MALE PLUMAGE

In all the species of ground-finch *Geospiza*, the juvenile male is coloured greyish brown like the female, but the fully plumaged male is wholly black except for white under the tail. The male first acquires the black feathering round the base of the beak, then on the rest of the head, then on the upper breast and back, and finally on the lower back and abdomen.

While many male *Geospiza* breed in fully black plumage, other individuals breed in plumage of juvenile pattern without any black, or in an intermediate stage with black confined to the head and breast, or simply to the head, or only to the front part of the head. These latter individuals have normally moulted out of their original juvenile plumage, but some or all of their new set of feathers are of juvenile (or female) pattern. Observations by Dr R. T. Orr on our captive finches show that the male normally develops black on the head in the second year, farther down the body in the third year, and all over in the fourth year, but there is much individual variation, so that these stages overlap.

Turning to the genus *Camarhynchus*, in the vegetarian tree-finch *C. crassirostris* and in the insectivorous tree-finches *C. psittacula*, *C. pauper*, and *C. parvulus*, the full male plumage is black on the head, upper breast and upper back, but not further down the body, i.e. the final stage in these species corresponds to a transitional stage in *Geospiza.* The black extends about half-way down the body in *Camarhynchus crassirostris* and about one-third of the way down in the other species. Actually two specimens of *C. crassirostris* and one of *C. psittacula* had black feathering over almost the whole body, showing that as a rarity these species attain the condition which in *Geospiza* is normal. In these four species of *Camarhynchus,* as in *Geospiza,* many individuals breed in plumage of juvenile pattern without black, or in an intermediate stage with black confined to the head or to just the front part of the head. Indeed, in the Chatham race of *C. parvulus,* most males remain permanently in plumage of juvenile pattern, and only a very few acquire black on head and breast. This leads up to the condition in the fifth species of *Camarhynchus,* the woodpecker-finch *C. pallidus,* in which the male normally shows no black at all, but out of hundreds seen in the field one had a black head, thus linking this species with the rest. Finally, in the mangrove-finch *C. heliobates,* black male plumage is unrecorded.

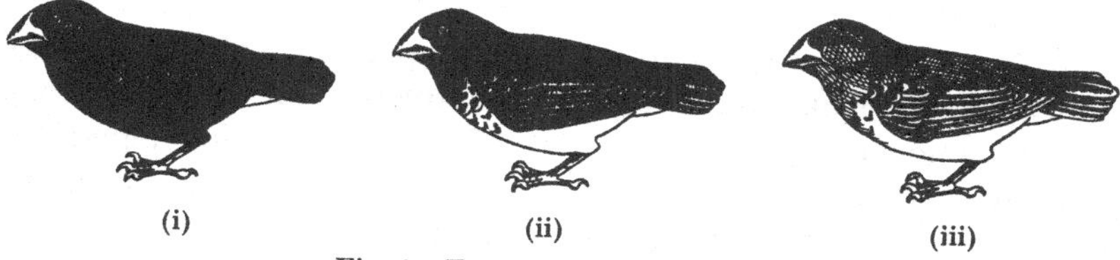

Fig. 8. Types of male plumage.

(i) Fully black male—final stage in *Geospiza.*
(ii) Partly black male—final stage in *Camarhynchus crassirostris, C. psittacula* and *C. parvulus,* transitional stage in *Geospiza.*
(iii) Plumage of juvenile pattern—permanent stage in *Camarhynchus pallidus,* initial stage in other species of *Camarhynchus* and in *Geospiza.*

Those individual males of *Geospiza* and *Camarhynchus* which breed in plumage of juvenile or intermediate pattern also show other characters suggesting immaturity. First, there is a general tendency in song-birds for the older males to start breeding ahead of the first-year males; in both *Geospiza* and *Camarhynchus* the males in black plumage tend to start nesting before those in plumage of juvenile pattern. Secondly, in many passerine birds the older males have on the average a longer wing than the first-year males; in Darwin's finches the average wing-length is greatest in fully black males, slightly smaller in partly black males, and a little smaller still in males with no black, as shown in Table XX, p. 169. Thirdly, in all of Darwin's finches the beak is coloured dark horn in the breeding season and pink outside the breeding season; fully plumaged males tend to acquire the dark beak earlier in the season than do those in plumage of juvenile or intermediate pattern, and the latter birds may even start to breed with an incompletely dark beak.

Experimental work has shown that in some birds the development of male plumage is primarily controlled by the sex hormones, whereas in others the control is primarily genetic. The factors involved in Darwin's finches are not known, and the situation is so unusual that an experimental investigation might yield results of general interest. Thus in *Geospiza* all stages up to the fully black condition are primarily dependent on age (and therefore under hormone control?) but in *Camarhynchus* the control is presumably rather different, as in most species the adults remain permanently half black, and only those stages up to the half-black condition depend primarily on age. Moreover, in both genera the age at which the final stage is acquired varies individually. Finally, there are other species of *Camarhynchus* in which the black male plumage is completely suppressed except as an abnormality. The concept of rate genes developed by Huxley (1942) might perhaps yield fruitful results, but more cannot be said until genetic and hormonal experiments have been carried out.

The above series also afford an example of neoteny, i.e. of the permanent retention into adult life of essentially juvenile characters. As already noted, the males which breed in plumage of juvenile pattern also show other features suggestive of immaturity, though the birds are fully mature sexually and raise normal families.

In the species of *Camarhynchus*, and perhaps also in *Geospiza*, the proportion of fully plumaged males is different on different islands. The proportion among collected males is set out in Tables XVIII and XIX, p. 168, and even making considerable allowance for biasing of the collected sample, it is clear that in *Camarhynchus* there are genuine island variations. The most striking case is in the small insectivorous tree-finch *C. parvulus*, in which of eighty-seven males collected on Charles 70 per cent are partly black, but of ninety-one collected on Chatham only 4 per cent are partly black. Other variations are apparent from Table XIX, and they are quite haphazard. For instance, on Charles the proportion of partly black males is unusually high in *C. parvulus*, but unusually low in both *C. crassirostris* and *C. pauper*. The figures for the species of *Geospiza* are suggestive of similar variations, but are inconclusive owing to the possible large sampling errors.

Although Darwin's finches display vigorously, neither the black plumage of the male *Geospiza*, nor the partly black plumage of the male *Camarhynchus* are specially displayed, and these colours do not emphasize the displays in any way. Further, those males breeding in plumage of juvenile type have a display indistinguishable from that of fully plumaged males of their species; and in the woodpecker-finch *C. pallidus*, in which the males are normally without black, the display is similar to that of the other species of *Camarhynchus*. It seems clear that the black male plumage of *Geospiza* and *Camarhynchus* has no function in display.

Another possibility is that it might assist the birds to discriminate the sex of strange individuals. But such discrimination plays no important part in the lives of Darwin's finches. At pair formation a potential mate is recognized primarily by its behaviour, and not by its plumage. Rivals of the same sex are also recognized primarily by their behaviour, and not by their plumage. The ground-finches are *capable* of distinguishing fully black males from those in plumage of juvenile or female type. Our field observations showed that the mated hen attacks all intruders in plumage of the latter type, but not those in fully black plumage. Also, when stuffed specimens of both types were placed at various nests, the wild cocks courted the stuffed female not infrequently, but the stuffed male only rarely (Lack, 1945). However, this ability to discriminate between the two types of plumage seems of no importance in the lives of the birds.

UNDER-TAIL COVERTS

In *Geospiza* the fully plumaged male is wholly black except for the under-tail coverts, which in most species are tipped with white, but in the sharp-beaked ground-finch *G. difficilis* are tipped with rufous.

The last holds for all the males of the Culpepper and Wenman form, for about half the males of the James and Indefatigable form, and for one-tenth of the males of the Tower and Abingdon form. These racial differences run parallel with the frequency of rufous in the wing-bar of the females in this species (see p. 41), and seem equally pointless. In black males of the Cocos-finch *Pinaroloxias*, the under-tail coverts are tipped with buff. So far as known, neither rufous, buff nor white under-tail coverts have any function in Darwin's finches, and in particular they are not used in display.

THE COCOS-FINCH

In the Cocos-finch *Pinaroloxias*, as in *Geospiza*, the full male plumage is wholly black, but the regulating factors are evidently somewhat different, since the juvenile male acquires the black feathering irregularly all over the body, instead of first at the head and then posteriorly. It is not known whether the Cocos-finch ever breeds in plumage of juvenile or transitional pattern.

THE WARBLER-FINCH

The male warbler-finch *Certhidea* is not black. Instead, the adult male is distinguished from the female by an orange-tawny patch on the lower part of the throat and the upper breast (see Plate VIII). Both the size of this patch, and the number of individuals showing it, differ markedly in different island forms. The patch is best developed on James, rather less so on Albemarle, and less again on Indefatigable. On Charles it is further reduced, but many males still possess it. On Culpepper and Wenman there are definite traces in all adult males, but the area of the patch is much smaller and its colour less bright. On Abingdon, Bindloe, Tower and Chatham, traces of the patch occur on some individuals only. In the birds of Barrington and Hood there is scarcely any trace at all. Hence in *Certhidea* the racial differences in the extent of the male plumage are even greater than in the other species of Darwin's finches.

The variations in the rufous throat patch of the male *Certhidea* tend to run parallel with those of the rufous wing-bar, present in both sexes (see Table VIII, p. 41). Wing-bar and throat patch are best developed on Albemarle, Indefatigable and James, less so on Charles, Culpepper and Wenman, and poorest on Hood and

Barrington. However, the correlation is not complete, since the Chatham race has a prominent rufous wing-bar, but the throat patch is poor, and there are other smaller discrepancies. On the whole, these rufous parts are most pronounced in the races with most olive in their general body-plumage, and are poorest in those with least olive. So far as observed, the throat patch does not feature in the display of the male *Certhidea*.

An orange-tawny throat patch is not normally found in any other species of Darwin's finches, but there are definite traces of it in one male specimen of each of four species, namely, *Geospiza fuliginosa*, *G. difficilis*, *Camarhynchus pauper* and *C. parvulus*. Such abnormal specimens provide another link between *Certhidea* and the other species.

LOSS OF DISTINCTIVE MALE PLUMAGE

Darwin's finches are highly unusual in that, though most species possess distinctive male plumage, many individual males breed without it, in plumage of juvenile pattern. The distinctive male plumage is not important in display, or in sex recognition; indeed, it appears to be without any function. All the evidence suggests that both the black plumage of *Geospiza* and *Camarhynchus* and the orange-tawny throat of *Certhidea* are useless characters in process of being lost. The loss is occurring faster in *Camarhynchus* than in *Geospiza*, and in some species it is occurring at a different rate on different islands. The alternative hypothesis, that distinctive male plumage is in process of being acquired by Darwin's finches, can be ruled out as too improbable, since this would mean supposing that many different forms are acquiring similar plumage independently.

In the vermilion flycatcher *Pyrocephalus rubinus*, a widespread bird in Central America, the male is a brilliant red. But in the Galapagos form of this species, only some of the males have vermilion plumage, and many breed in dull plumage of juvenile pattern. The latter males display as vigorously as those in full plumage, as is the case in Darwin's finches. The Galapagos martin *Progne modesta* is also reported to breed in plumage of juvenile pattern (Beebe, 1924).

In general, there is a marked tendency for the land birds of oceanic islands to lose distinctive male plumage. In the European bullfinch *Pyrrhula pyrrhula*, the adult male is bright pink below, the female and juvenile are grey, but the adult male of the Azorean bullfinch *P. p. murina* is grey below, like the juvenile (Murphy and Chapin, 1929). The Hawaiian flycatcher *Chasiempsis* not infrequently breeds in plumage of juvenile type (Perkins, 1903), and the same is true of the finch *Nesospiza* on Tristan da Cunha (Lowe, 1923). Many similar instances are given by Mayr (1931–4) for the land birds of Polynesian islands. In some of these, such as the cuckoo-shrike *Coracina lineata* and the flycatchers *Pachycephala pectoralis* and *Petroica multicolor*, there is, as in Darwin's finches, inter-island variation, some island races showing loss of male plumage and others not. Another parallel with *Geospiza* is provided by the flycatchers *Pomarea iphis* and *Myiagra vanikorensis*, in which some males have plumage intermediate in pattern between that of the juvenile and the fully plumaged male.

Examples of the same phenomenon can also be found in continental birds, the best known being that of the common or red crossbill *Loxia curvirostra*, in which many males breed in greenish plumage like that of the juvenile, and the proportion of red males in the population varies with the geographical race (Griscom, 1937). But the loss of bright male plumage is proportionately much commoner in the song-birds of oceanic islands than in mainland forms. There is no apparent reason why courtship should be less important on oceanic islands than on the continents, and in any case birds such as Darwin's finches and the Galapagos vermilion flycatcher display quite as vigorously as mainland birds. But there is another function, sometimes overlooked, for bright male plumage. This plumage is highly distinctive for each species, and therefore it assists the female to recognize and pair with a male of the correct species. Interbreeding between different species is at a selective disadvantage, particularly as the offspring are often sterile, and hence specific recognition marks have survival value. However, owing to the difficulties of colonization, the species of land-birds on oceanic islands are few in number, and they are usually away from all related species with which a hybrid pair might possibly be formed. As a result, the male's specific recognition marks become largely unnecessary. It is suggested that this is probably the main reason why such birds tend to lose their distinctive plumage.

In Darwin's finches the position is more complex than in the other cases cited, since the original form has later diverged into a number of species, and so need has again arisen for the female to distinguish accurately a male of her own kind. It is difficult to see why, in this case, specific plumage characters have not been evolved afresh, since all that would be needed is stabilization of the progressive loss of black male plumage at a different stage in each species. But this has not occurred, and in these birds specific recognition now depends on characters other than plumage, as discussed below.

SPECIFIC RECOGNITION

In Darwin's finches each individual normally pairs with one of its own species, and each defends its territory primarily against intruders of its own species. Clearly the birds can distinguish members of their own kind, and field observations on attacking behaviour revealed the chief recognition factor involved. On seeing a strange finch in its territory, the owning bird usually starts its attack by flying down to it, then comes round in front to face it. If the intruder is of its own species, attack follows, but if it belongs to another species, the owner's aggressive behaviour usually peters out, i.e. it disappears at the moment when it sees the beak of the stranger. This was observed in the large ground-finch *Geospiza magnirostris* when starting to attack a medium ground-finch *G. fortis*; in *G. fortis* when starting to attack a small ground-finch *G. fuliginosa*, also against a woodpecker-finch *Camarhynchus pallidus*; in *Geospiza fuliginosa* against *G. fortis*; and in the large insectivorous tree-finch *Camarhynchus psittacula* against the small species *C. parvulus*. It was observed so often that it became clear that in Darwin's finches the beak is the chief character used in specific recognition.

This conclusion was checked by field experiments. Specimens of the small and medium ground-finches *Geospiza fuliginosa* and *G. fortis* were stuffed so that they differed little in body size, and since their plumage is identical, the difference between them was reduced effectively to a difference in beak. These specimens were set up at nests of the small ground-finch *G. fuliginosa* in the wild.

When the stuffed female specimen of their own species was placed near the nest, three wild male *G. fuliginosa* courted it.

When the stuffed female *G. fortis* was substituted, one male ceased to react and the other two reacted only feebly. When the stuffed *G. fuliginosa* was returned, all three males courted it vigorously. Clearly, they distinguished the specimen of their own species. That in two cases discrimination was not perfect may be attributable to the artificial conditions of the experiment.

Similarly, when the stuffed female of their own species was placed near the nest, two mated female *G. fuliginosa* attacked it vigorously. When the stuffed female *G. fortis* was substituted, one bird ceased to attack and the other attacked only feebly. When the stuffed female *G. fuliginosa* was returned, both birds again attacked it vigorously. Clearly they distinguished the specimen of their own species. Partial confirmation was received from a third wild female *G. fuliginosa*, which vigorously attacked a stuffed female of its own species, then when presented with a stuffed female *G. fortis* showed no aggressive behaviour; but thereafter it lost all interest in stuffed specimens, as sometimes happens.

These and other experiments with stuffed specimens, and the difficulties in interpreting such work, are discussed in detail elsewhere (Lack, 1945). The above six cases all indicate that the small ground-finch *G. fuliginosa* can correctly distinguish members of its own species from those of the larger *G. fortis* by means of beak differences, and no observations or experiments suggested the contrary.

On consideration, it is not surprising that Darwin's finches should recognize their own kind primarily by beak characters. The beak is the only prominent specific distinction, and it features conspicuously both in attacking behaviour, when the birds face each other and grip beaks, and also in courtship, when food is passed from the beak of the male to the beak of the female. Hence though the beak differences are primarily correlated with differences in food, secondarily they serve as specific recognition marks, and the birds have evolved behaviour patterns to this end.

In Darwin's finches the beak serves the same function as the wing-bars, head markings or tail patterns which so often distinguish closely related species in other birds. It is the frequency with which visual recognition marks have been evolved in birds that makes their systematics so easy, for the museum worker often distinguishes closely related species by the features specially evolved for this purpose in the birds themselves. Correct identification is much more difficult in groups such as the leaf warblers *Phylloscopus* (Ticehurst, 1938) and the grass warblers *Cisticola* (Lynes, 1930), in which the females appear to recognize males of their own species primarily by vocal characters. In Darwin's finches the species are not readily distinguishable either by plumage or by voice, and the beak (which, after all, has another and more important function) is somewhat imperfect as a recognition mark, since in some species it is subject to marked individual variation. Possibly Darwin's finches have diverged from each other so recently that recognition marks are as yet incompletely evolved; nevertheless, interbreeding between species seems to be extremely rare.

Chapter VI: BEAK DIFFERENCES AND FOOD

The most curious fact is the perfect gradation in the size of the beaks in the different species of Geospiza, from one as large as that of a hawfinch to that of a chaffinch, and (if Mr Gould is right in including his sub-group, Certhidea, in the main group), even to that of a warbler....The beak of Cactornis is somewhat like that of a starling; and that of the fourth sub-group, Camarhynchus, is slightly parrot-shaped.

Charles Darwin: *The Voyage of the 'Beagle'*, Ch. xvii.

SUBGENERA

The chief way in which the various species of Darwin's finches differ from each other is in their beaks. Indeed, the beak differences are so pronounced that systematists have at various times used as many as seven different generic names for the birds. In this book the genera are reduced to four, but it is convenient to retain the other generic names as subgenera, since they emphasize the adaptive radiation of the finches, as set out in Table X. Table X refers only to the species on the central Galapagos islands, the forms on outlying islands being considered later.

The observations of Gifford (1919) and ourselves show that the marked beak differences between the subgenera of Darwin's finches are correlated with marked differences in feeding habits.

(i) Subgenus *Geospiza*. The heavy finch-like beaks of the large, medium and small ground-finches *Geospiza magnirostris*, *G. fortis* and *G. fuliginosa* suggest a diet of seeds, and this is the chief food of these birds, particularly outside the breeding season. In addition, they eat various fruits, also flowers, buds, young leaves and large caterpillars at the seasons when these are particularly abundant. Small insects and spiders are taken occasionally.

Few observations are available on the food of the fourth species, the sharp-beaked ground-finch *G. difficilis*. The race *G. d. debilirostris* of the central Galapagos islands feeds on the ground in the humid forest, scratching about in the leaves (Gifford, 1919), thus occupying the niche filled in English woods by the blackbird *Turdus merula*. The beak of *Geospiza d. debilirostris* fits this habit, since it is rather more elongated than that of typical ground-finches such as *G. fortis* and *G. fuliginosa*.

TABLE X. ADAPTIVE RADIATION ON CENTRAL GALAPAGOS ISLANDS

	Subgenus	Species	Beak	Chief food	Comments
1.	*Geospiza*	*magnirostris* *fortis* *fuliginosa* *difficilis*	Heavy, finch-like	Seeds	Four ground-finche three in coast zone differing size, one in hum zone
2.	*Cactornis*	*scandens*	Long, decurved	*Opuntia*	Cactus feeding
3.	*Platyspiza*	*crassirostris*	Thick, short and somewhat decurved	Buds, leaves, fruit	Primarily vegetari
4.	*Camarhynchus*	*psittacula* *pauper* *parvulus*	Like the last	Insects	Two species (thr on Charles), diffe ing in size
5.	*Cactospiza*	*pallidus* *heliobates*	Stout, straight	Insects, especially from wood	Two species, diff ing in habitat
6.	*Certhidea*	*olivacea*	Slender	Small insects	Warbler-finch

Notes. (i) *Cactornis* is part of the genus *Geospiza*, and *Platyspiza* and *Cactospiza* are part of the genus *Camarhynchus*.

(ii) The seventh subgroup is *Pinaroloxias* of Cocos, with a slender beak and insectivorous habits. There is only one species.

(ii) Subgenus *Cactornis*. The long, somewhat decurved beak and the split tongue of the cactus ground-finch *Geospiza scandens* suggest a flower-probing and nectar-feeding habit, and the flowers of the prickly pear *Opuntia* are a staple food of this species. It also feeds regularly on the soft pulp of the prickly pear, and on all the types of food, including seeds, listed above as taken by the other ground-finches. Hence, though partly specialized in feeding habits, it has not completely departed from the usual ground-finch diet, and, probably correlated with this, its beak is thicker than that of typical flower-eating birds. Gifford (1919) recorded that on Charles *Geospiza scandens* has taken to feeding on the fruits of the introduced oranges and tropical plums, an interesting modification of its natural feeding habits.

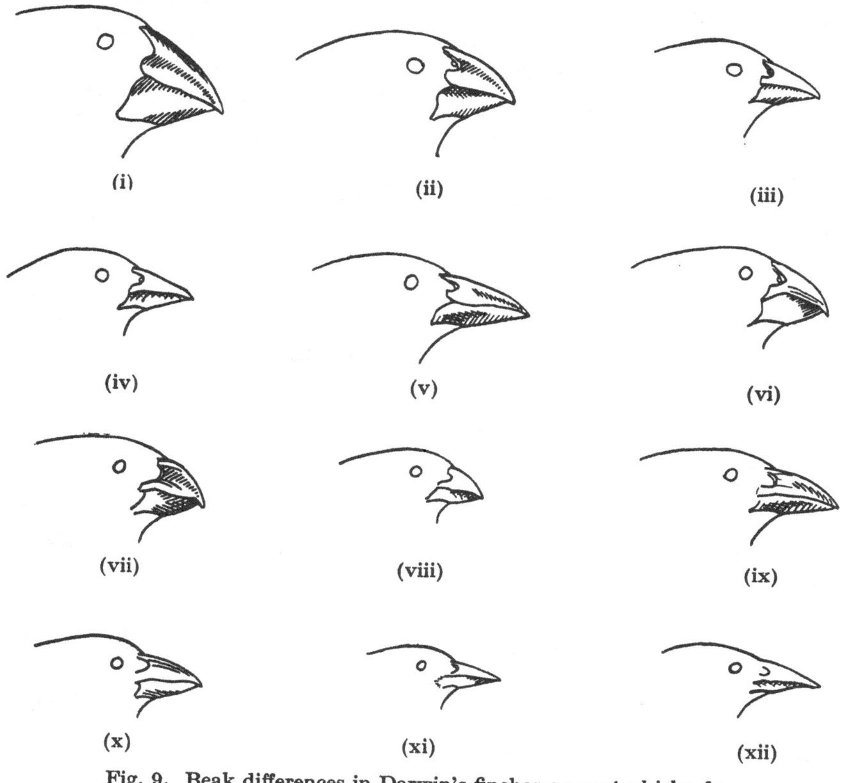

Fig. 9. Beak differences in Darwin's finches on central islands. $\frac{2}{3}$ natural size (*after* Swarth).

(i) *Geospiza magnirostris* (ii) *Geospiza fortis* (iii) *Geospiza fuliginosa*
(iv) *Geospiza difficilis debilirostris* (v) *Geospiza scandens* (vi) *Camarhynchus crassirostris*
(vii) *Camarhynchus psittacula* (viii) *Camarhynchus parvulus* (ix) *Camarhynchus pallidus*
(x) *Camarhynchus heliobates* (xi) *Certhidea olivacea* (xii) *Pinaroloxias inornata*

(iii) Subgenus *Platyspiza*. The vegetarian tree-finch *Camarhynchus crassirostris* has a short and thick, somewhat decurved and slightly parrot-like beak, rather similar in appearance to that of other birds which, like it, feed primarily on leaves, buds, blossoms and fruits. This species rarely takes insects, and was not seen to eat grain, but it takes some types of seeds. It feeds mainly in the trees, but comes to the ground to take young leaves of herbaceous plants and fallen fruit. Its actions are leisurely, like those of many other bud- and fruit-eating birds.

(iv) Subgenus *Camarhynchus*. The insectivorous tree-finches *Camarhynchus psittacula*, *C. pauper* and *C. parvulus* have beaks very similar in shape to that of the vegetarian tree-finch *C. crassirostris*. This resemblance is presumably due to close relationship rather than to adaptive modification, since the food of the three former species is quite different from that of *C. crassirostris*, consisting mainly of beetles and similar insects, for which the birds examine the twigs, bark and leaf clusters, and also excavate shallow holes in soft wood. They feed chiefly in the trees and are agile in their movements, rather like tits, sometimes turning almost upside down in their search for food. They feed on the ground at times. While moderately small insects form their main food, they also eat nectar, buds, young leaves and large caterpillars at the seasons when these are abundant, and occasionally take grain.

(v) Subgenus *Cactospiza*. The woodpecker-finch *Camarhynchus pallidus* has a stout, straight beak, with obvious affinities to that of the insectivorous tree-finches, but more elongated, and modified in the direction of that of a woodpecker or nuthatch. It feeds on beetles and similar insects, for which it searches bark and leaf clusters, and less commonly the ground, and also bores into wood. It is much more exclusively insectivorous than the insectivorous tree-finches, and with this can be correlated the greater specialization of its beak.

C. pallidus further resembles a woodpecker in that it climbs up and down vertical trunks and branches. It is the only one of Darwin's finches to do this. It also possesses a remarkable, indeed a unique, habit. When a woodpecker has excavated in a branch for an insect, it inserts its long tongue into the crack to get the insect out. *C. pallidus* lacks the long tongue, but achieves the same result in a different way. Having excavated, it picks up a cactus spine or twig, one or two inches long, and holding it lengthwise in its beak, pokes it up the crack, dropping the twig to seize the insect as it emerges. In the arid zone the bird uses one of the rigid spines of the prickly pear *Opuntia*, but in the humid zone, where there is no *Opuntia*, it breaks off a small twig of suitable length from a tree or bush. It has been seen to reject a twig if it proved too short or too pliable. Sometimes the bird carries a spine or twig about with it, poking it into cracks and crannies as it searches one tree after another. This remarkable habit, first reported by Gifford (1919) and fully confirmed first by

Fig. 10. *Camarhynchus pallidus* and its stick.
(Drawn by Roland Green from photographs by R. Leacock.)

W. H. Thompson and later by the writer, is one of the few recorded uses of tools in birds. The nearest parallel is the use of fruits by the bower-bird *Ptilonorhynchus violaceus* for staining the stems in its bower (Gilbert, 1939).

The mangrove-finch *Camarhynchus heliobates* has a beak similar to that of *C. pallidus* and, like the latter, feeds almost exclusively on insects (Snodgrass, 1902*a*). Its feeding methods have not been recorded.

(vi) Genus *Certhidea*. The beak, general habits and appearance of the warbler-finch *Certhidea* are so like those of a warbler that it was for a long time considered to be one. Like a warbler, it searches the leaves, twigs and ground vegetation for small insects, and sometimes catches insects in the air. Nectar from flowers and young green leaves are also taken in season, but small soft insects form the great majority of its diet. The Hood form of this species also feeds commonly on small marine organisms collected at low tide below high-water mark, but this habit has rarely been observed on other islands, so provides an example of a racial difference in feeding habits.

(vii) Genus *Pinaroloxias*. The Cocos-finch *Pinaroloxias* is said to feed predominantly on insects and has a slender beak similar to that of *Certhidea*, except that it is a little longer and somewhat decurved. *Pinaroloxias* feeds both on the ground and in the trees, and has an unusually long, grooved and bifid tongue (Snodgrass, 1903; Gifford, 1919).

To summarize, the beak differences between most of the genera and subgenera of Darwin's finches are clearly correlated with differences in feeding methods. This is well borne out by the heavy, finch-like beak of the seed-eating *Geospiza*, the long beak of the flower-probing *Cactornis*, the somewhat parrot-like beak of the leaf-, bud- and fruit-eating *Platyspiza*, the woodpecker-like beak of the woodboring *Cactospiza*, and the warbler-like beaks of the insect-eating *Certhidea* and *Pinaroloxias*. Only in one group, namely, the insectivorous tree-finches of the subgenus *Camarhynchus* (*sens. strict.*), is the beak not particularly suggestive of the feeding habits; these birds, though feeding primarily on insects, may be regarded as moderately unspecialized in both diet and beak.

CLOSELY RELATED SPECIES

While the beak differences between most of the subgenera of Darwin's finches are clearly adapted to differences in feeding methods, the same does not seem to hold for the beak differences between closely related species. Here the differences are chiefly in the size rather than the shape of the beak, and the species concerned have almost identical feeding methods. For instance, the large ground-finch *Geospiza magnirostris* has a huge beak, the medium *G. fortis* a fairly large one, and the small *G. fuliginosa* a smaller one. But all three species feed in similar places and have similar feeding methods, while their foods, listed in the preceding section, are of the same general nature. Similarly, the large insectivorous tree-finch *Camarhynchus psittacula* has a much larger beak than the small *C. parvulus*, but it occupies the same habitat and has almost identical feeding methods. The parallel with *Geospiza* is completed on Charles, where there occurs a third species, *Camarhynchus pauper*, intermediate in size and size of beak between the other two.

The significance of these marked beak differences between species otherwise similar has excited speculation from all who have discussed Darwin's finches. Our field observations confirm the similarity in feeding methods of the three species of *Geospiza* and that they often feed on the same things. However, we observed that the large, hard fruits of the manzanilla *Hippomane mancinella* are taken freely by the large *Geospiza magnirostris* and the medium *G. fortis*, but rarely if at all by the small *G. fuliginosa*; on the other hand, certain small seeds, notably of grasses, form a staple food of the small *G. fuliginosa*, but are taken less commonly by the medium *G. fortis*, and rarely if at all by the large *G. magnirostris*. Similarly, Snodgrass (1902*a*) showed from an analysis of stomach contents that the foods of the various species of *Geospiza* are often identical. However, the large *G. magnirostris* eats certain large seeds not taken by the small *G. fuliginosa*, while a series of the medium *G. fortis* and the small *G. fuliginosa* collected at the same time and place on North Albemarle were eating partly the same but partly different foods. Snodgrass was chiefly impressed by the similarity in the foods of the different species, but care is needed in interpreting his data, as too few stomach contents were collected to be truly representative of birds with such varied diets. In one case his results are definitely misleading, since they suggest that the small ground-finch *G. fuliginosa* and the cactus ground-finch *G. scandens* eat largely the same types of food. This happens to be true in the period immediately following the breeding season, when Snodgrass collected his specimens, for then *G. scandens*, like *G. fuliginosa*, eats many small seeds. But in the breeding season the diets of these two species are mainly different, *G. scandens* depending largely on prickly pear, as already noted.

Snodgrass concluded that the beak differences between the species of *Geospiza* are not of adaptive significance in regard to food. The larger species tend to eat rather larger seeds, but this he considered to be an incidental result of the difference in the size of their beaks. This conclusion was accepted by Gifford (1919), Gulick (1932), Swarth (1934) and formerly by myself (Lack, 1945). Moreover, the discovery mentioned in the last chapter, that the beak differences serve as recognition marks, provided a quite different reason for their existence, and thus strengthened the view that any associated differences in diet are purely incidental and of no particular importance.

My views have now completely changed, through appreciating the force of Gause's contention that two species with similar ecology cannot live in the same region (Gause, 1934). This is a simple consequence of natural selection. If two species of birds occur together in the same habitat in the same region, eat the same types of food and have the same other ecological requirements, then they should compete with each other, and since the chance of their being equally well adapted is negligible, one of them should eliminate the other completely. Nevertheless, three species of ground-finch live together in the same habitat on the same Galapagos islands, and this also applies to two species of insectivorous tree-finch. There must be some factor which prevents these species from effectively competing.

The above considerations led me to make a general survey of the ecology of passerine birds (Lack, 1944*a*). This has shown that, while most closely related species occupy separate habitats or regions, those that occur together in the same habitat tend to differ from each other in feeding habits and frequently also in size, including size of beak. In a number of the latter cases it is known that the beak difference is associated with a difference in diet, and this correlation seems likely to be general, since it is difficult to see how otherwise such species could avoid competing. It seems particularly significant that when two closely related species differ from each other in size and size of beak they often live in the same habitat, while such marked size differences are unusual in closely related species which occupy different habitats.

The general survey indicates that when two closely related species live in the same habitat, they do not usually take completely different foods. It is apparently sufficient if some of their foods are different, but the extent to which their diets can overlap is not known. In the case of the three species of *Geospiza*, there are similarities, but also established differences, in their diets, and though further evidence is much needed, it is provisionally concluded here that, so far from being unimportant and purely incidental, these food differences are essential to the survival of the three species in the same habitat. Further food analyses might most profitably be made in the latter part of the dry season, as this is the period when food is likely to be least varied and least abundant, and therefore most likely to limit the population density of the birds.

The food of the insectivorous tree-finches has not been analysed, but I suggest that, in this case also, the larger species probably tends to eat larger insects and the smaller species smaller insects, and that as a result the two species to some extent share out, instead of competing for, the available food supply. This view receives confirmation from evidence of a different nature. On Chatham, the large *Camarhynchus psittacula* does not occur, so it might be expected that a greater range of foods would thereby become available for the small species *C. parvulus*. It is therefore significant that the Chatham form of *C. parvulus* has a larger beak than any other island form of this species, indeed the beak overlaps in measurements with that of *C. psittacula* on other islands. This is just what would be expected if the beak difference between the two species is an adaptation for taking foods of different size, since on islands where *C. psittacula* is absent, but not elsewhere, unusually large individuals of *C. parvulus* will have survival value.

There is a parallel in the ground-finches, since the large *Geospiza magnirostris* is absent from Chatham and Charles, and on both these islands the medium *G. fortis* reaches a much greater maximum size than it does on the northern Galapagos islands, where *G. magnirostris* is common. Indeed, some of the *G. fortis* from Chatham and Charles are as large as small specimens of *G. magnirostris* from the northern islands. This point is discussed further in Chapter VIII.

To sum up, though the three species of *Geospiza* have similar feeding methods, I consider that the marked difference in the size of their beaks is an adaptation for taking foods of different size and that, superficial appearances to the contrary, these three species are food specialists to an extent sufficient to enable them to live in the same habitat without effectively competing. A similar view is advocated for the difference in size of beak between *Camarhynchus psittacula* and *C. parvulus*. However, more detailed analysis of the foods taken by these species is extremely desirable, particularly in the case of *Camarhynchus*.

SIZE DIFFERENCES IN OTHER BIRDS

Instances of closely related species which differ markedly in size and size of beak are fairly common among the land birds of other oceanic islands. For example, on Lord Howe Island, off Australia, occur two species of the white-eye *Zosterops* which differ markedly in size and size of beak, while on adjacent Norfolk Island there are three species, a large, a medium and a small (Mathews, 1928; Streseman, 1931), thus recalling the three species of *Geospiza* in the Galapagos. Again, on Nightingale Island in the Tristan da Cunha group, there are two species of the finch *Nesospiza*, differing primarily in size and size of beak (Lowe, 1923). The latter birds, together with the two species of *Zosterops* on Lord Howe, are shown in Fig. 24, p. 139. Two parallel examples are found on Kauai in the Hawaiian archipelago, where there are a large and a small species of the thrush *Phaeornis* and also of the sicklebill *Chlorodrepanis* (Perkins, 1903). Other instances are given elsewhere (Lack, 1944*a*).

Cases are also not uncommon in continental birds. In Europe, the great and small reed warblers *Acrocephalus arundinaceus* and *A. scirpaceus* breed side by side in reed beds, while the great, middle and little spotted woodpeckers, *Dryobates major*, *D. medius* and *D. minor*, provide another example of three species which live together in much the same habitat and differ primarily in size. In South America two species of the rice grosbeak *Oryzoborus* differ in size of beak as much as the large and medium ground-finches of the Galapagos. Again, a large and small species of babbler, *Garrulax pectoralis* and *G. moniliger*, occur together through Burma and Indo-China (Mayr, 1942). There are equally good examples in groups other than the song-birds, for instance in birds of prey, grebes, gulls, terns and many others.

Huxley (1942), who gives further examples, is apparently the only previous worker to suggest that these differences in size are primarily correlated with differences in food. He instances the case of the two British falcons, the small merlin *Falco columbarius* and the larger peregrine *F. peregrinus*, the former hunting small birds and the latter feeding mainly on large birds. He might perhaps have added a third species, the kestrel *F. tinnunculus*, which is intermediate in size and takes different prey from either of the other two, chiefly small rodents, but in this case there is also a difference in hunting methods. Again, three species of terns breed in the same localities on the east coast of England, the little tern *Sterna albifrons*, the medium-sized common tern *S. hirundo*, and the large sandwich tern *S. sandvicensis*. Huxley quotes from Gause (1934) some observations at a colony of these birds on the Black Sea which show that, though the three species all feed at sea and have similar feeding methods, they feed at different distances from the shore, presumably therefore on different kinds or sizes of food, and so do not compete. In the Black Sea ternery there also occurs a fourth species, the gull-billed tern *Gelochelidon nilotica*, which feeds inland on land insects.

Similarly, on Kauai in Hawaii, Perkins (1903) found a difference in feeding habits between the small and the large species of the thrush *Phaeornis* and between the small and the large species of the sicklebill *Chlorodrepanis*. Likewise, Hagen (1940) reports that the larger species of *Nesospiza* on Nightingale Island eats nuts in trees, whereas the smaller species feeds in the tussock grass. Again, the English countryside supports two species of pigeon, the larger ring dove *Columba palumbus* and the smaller stock dove *C. oenas*, and M. K. Colquhoun informs me that the investigation by the British Trust for Ornithology has shown that these two species eat mainly different foods. A particularly clear example is provided by the crossbills of northern Europe, since the small-beaked two-barred crossbill *Loxia leucoptera* feeds chiefly on the soft cones of the larch, the medium-sized common crossbill *L. curvirostra* on spruce cones, and the heavy-beaked parrot crossbill *L. pytyopsittacus* on the hard cones of the pine. When the common crossbill does take pine cones, it deals with them less efficiently than does the parrot crossbill (Niethammer, 1937; Lack, 1944*a*, *b*).

If more were known about the food of birds, the above examples could probably be multiplied, but they are sufficient to illustrate the view advanced in the preceding section that, in general, when closely related species differ markedly in size of beak, they also tend to differ in diet. The significance of the fact that such cases occur chiefly in species which live in the same habitat has already been remarked.

GEOSPIZA ON THE REMOTE ISLANDS

On the small, remote and low-lying islands of Culpepper, Wenman and Tower in the north and Hood in the south-east, the ecological picture is rather different from that on the central Galapagos islands. Whereas in the arid zone of the central islands there are normally four species of *Geospiza*, on these remote islands both the medium ground-finch *G. fortis* and the cactus ground-finch *G. scandens* are invariably absent, while the large ground-finch *G. magnirostris* and the small *G. fuliginosa* are sometimes missing. In the absence of one of these species, the vacant ecological niche may be filled by a different species, while when two of these species are absent, one form sometimes occupies two niches. The beak modifications of the birds concerned provide an interesting parallel with those of the ground-finches on the central islands, and lend further support to the view that such modifications are adapted to the nature of the food. The ecological picture is summarized in Table XI.

TABLE XI. ECOLOGICAL NICHES ON OUTLYING ISLANDS

Niche	Tower	Hood	Wenman	Culpepper
Large ground-finch	*G. magnirostris*	↑ *G. conirostris*	*G. magnirostris*	*G. conirostris*
Cactus-feeder	*G. conirostris*	↓	↑ *G. difficilis*	↑ *G. difficilis*
Small ground-finch	*G. difficilis*	*G. fuliginosa*	↓	↓
Warbler-finch	*Certhidea*	*Certhidea*	*Certhidea*	*Certhidea*

On Tower, as on the central Galapagos islands, the large-ground-finch niche is filled by *G. magnirostris*. But the usual small ground-finch *G. fuliginosa* is missing from Tower, and this niche is here occupied by the sharp-beaked ground-finch *G. difficilis*, a species which on the central islands feeds on the floor of the humid forest. In beak and size, the Tower form of *G. difficilis* is extremely similar to *G. fuliginosa*, a resemblance attributable to parallel evolution. The usual cactus ground-finch *G. scandens* is also absent from Tower; its place is taken by the species *G. conirostris*, the Tower form of which has a deeper beak but is otherwise very similar to typical *G. scandens*, so that it should perhaps be regarded simply as a highly modified race of *G. scandens*. The deeper beak of the Tower *G. conirostris* suggests that, in addition to feeding on cactus, this form perhaps eats some of the foods normally taken by the medium ground-finch *G. fortis*, which is absent from Tower.

Coming now to Hood, the small-ground-finch niche is filled as on the central islands by *G. fuliginosa*. But neither the large ground-finch *G. magnirostris*, nor the medium *G. fortis*, nor the cactus ground-finch *G. scandens* occur on Hood, and these niches are all filled by one form, namely, the Hood race of *G. conirostris*. The latter feeds both on the ground on seeds and also on prickly pear *Opuntia*, and its beak is intermediate in appearance between that of the ground-feeding *G. magnirostris* and *G. fortis* on the one hand, and that of cactus-feeding forms like *G. scandens* and Tower *G. conirostris* on the other. This Hood form is basically closest to *G. conirostris*, to which species it clearly belongs, and its superficial resemblance in beak to *G. magnirostris* is attributable to parallel evolution due to similar feeding habits.

On Wenman, the situation is again different. The large-ground-finch niche is filled in the usual way by *G. magnirostris*, but neither the small ground-finch *G. fuliginosa* nor the cactus ground-finch *G. scandens* occurs, and these two niches are occupied by the Wenman form of the sharp-beaked ground-finch *G. difficilis*. The latter feeds regularly both on the ground and on *Opuntia* (Rothschild and Hartert, 1899; Gifford, 1919), and its beak is larger and longer than that of the other races of *G. difficilis*, which are not cactus-feeders. Indeed, its beak is sufficiently like that of the true cactus ground-finch *G. scandens* for some systematists to have regarded the Wenman bird as a race of *G. scandens*. However, plumage and other characters prove that it is really a form of *G. difficilis*.

On Culpepper, which is close to Wenman, the same form of *G. difficilis* again combines the rôles of small ground-finch and cactus-feeder, and has the same type of beak. The large ground-finch *G. magnirostris* is absent from Culpepper, and this niche is

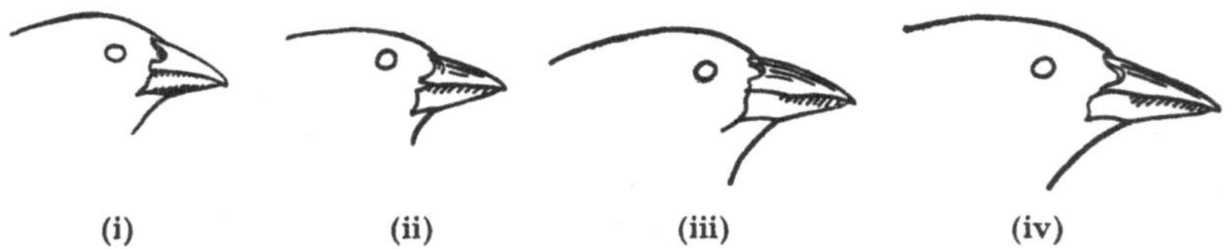

Fig. 11. Beak adaptations in *Geospiza difficilis*.
$\frac{2}{3}$ natural size (*after* Swarth).
(i) *G. fuliginosa* (ii) *G. difficilis difficilis* (iii) *G. difficilis septentrionalis* (iv) *G. scandens* (James form)

TABLE XII. BEAK ADAPTATIONS IN *GEOSPIZA DIFFICILIS*

Species	Island	Niche	Average measurement in mm. Whole culmen	Average measurement in mm. Depth of beak
G. fuliginosa	Hood	Typical small ground-finch	11·9	8·3
G. difficilis difficilis	Tower	Small ground-finch	12·4	7·9
G. difficilis septentrionalis	Culpepper	Small ground-finch and cactus-feeder	15·5	9·0
G. scandens	James	Typical cactus-feeder	17·0	8·8

Note. The measurements for the whole culmen are taken from Swarth (1931), those for the depth of beak are my own. For numbers measured see Supplementary Table C, p. 184.

here filled by the form *G. conirostris darwini*. The beak of the latter shows a strong basic resemblance to that of the other races of *G. conirostris*, but it is deeper and stronger, superficially like that of *G. magnirostris*. The latter resemblance is probably due to parallel evolution, as it is the type of beak which would be expected in a form of *G. conirostris* which had become specialized primarily for ground feeding. (Another possibility, that *darwini* is of hybrid origin between *G. conirostris* and *G. magnirostris*, cannot be altogether ruled out, but it seems very improbable; see p. 97.)

To summarize, on one outlying island the species *G. difficilis* occupies the small-ground-finch niche, while on another it combines this with cactus-feeding. Similarly, on one island the

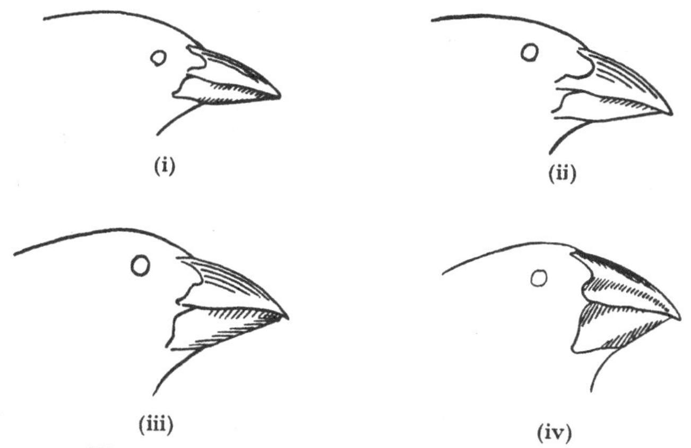

Fig. 12. Beak adaptations in *Geospiza conirostris*. ⅔ natural size (*after* Swarth).
(i) *G. scandens* (Bindloe) (ii) *G. conirostris propinqua* (Tower)
(iii) *G. conirostris conirostris* (Hood) (iv) *G. magnirostris* (Tower)

TABLE XIII. BEAK ADAPTATIONS IN *GEOSPIZA CONIROSTRIS*

Species	Island	Niche	Average measurement in mm. Whole culmen	Depth of beak
G. scandens	Bindloe	Typical cactus-feeder	19·3	10·6
G. conirostris propinqua	Tower	Cactus-feeder	19·5	13·0
G. conirostris conirostris	Hood	Large ground-finch and cactus-feeder	21·3	16·0
G. conirostris darwini	Culpepper	Large ground-finch	21·5	16·5
G. magnirostris	Tower	Typical large ground-finch	23·9	21·2

Note. The measurements for the whole culmen are taken from Swarth (1931), those for the depth of beak are my own. For numbers measured see Supplementary Table C, p. 184.

species *G. conirostris* combines the rôles of large ground-finch and cactus-feeder, on another it is predominantly cactus-feeding and on another it is primarily a large ground-finch. In each case, the beak of the form concerned is correspondingly modified. These modifications are shown in Figs. 11 and 12 and Tables XII and XIII.

Fifteen specimens of the medium ground-finch *G. fortis* have been collected on Hood and one on Wenman, and the cactus ground-finch *G. scandens* has been observed at sea 20 miles from land (see p. 21). Such occurrences suggest that the normal absence of these two species from the outlying Galapagos islands is due not to their inability to reach them, but to the fact that, when they reach them, they fail to establish themselves there. Such failure might be because, on a small island, comparatively specialized feeders like *G. scandens* and *G. fortis* survive less well than species which fill wider ecological niches, like *G. conirostris* and the Wenman form of *G. difficilis*.

CAMARHYNCHUS ON OUTLYING ISLANDS

In discussing the ecological niches of the large and small insectivorous tree-finches *Camarhynchus psittacula* and *C. parvulus*, a third species, the woodpecker-finch *C. pallidus*, should also be considered, since all three species feed in a similar way on similar types of food. *C. pallidus* is about the same size as *C. psittacula*, but whereas the beak of the latter is short, thick and decurved, that of *C. pallidus* is longer and straight, more specialized for excavating in wood. *C. pallidus* is further specialized for climbing vertical trunks, and is also able to probe insects out of crevices, which *C. psittacula* cannot do. Presumably the foods of *C. psittacula* and *C. pallidus*, though in part similar, are sufficiently different to enable the two species to live together in the same habitat without effectively competing.

On James and Indefatigable all three species occur together. But on Chatham the large insectivorous tree-finch *C. psittacula* is absent, and here, as already mentioned, the small species *C. parvulus* attains an unusually large size. But this is not all, for the Chatham form of the woodpecker-finch *C. pallidus* has a shorter beak than usual. This suggests that, in the absence of *C. psittacula*, there has been survival value to *C. pallidus* in becoming less specialized in beak, and that the Chatham form of *C. pallidus* takes some of the foods which on other islands are taken by *C. psittacula*.

The opposite situation perhaps occurs on Abingdon and Bindloe, where the woodpecker-finch *C. pallidus* is absent, and the large insectivorous tree-finch *C. psittacula* has a longer and straighter beak than usual, suggesting that it may take some of the foods normally taken by *C. pallidus*. But the resemblance in beak to *C. pallidus* is not close, so is perhaps fortuitous. On Bindloe the small insectivorous tree-finch *C. parvulus* is also absent, but as the Bindloe form of *C. psittacula* is no smaller than usual, the *C. parvulus* niche is presumably unoccupied on this island.

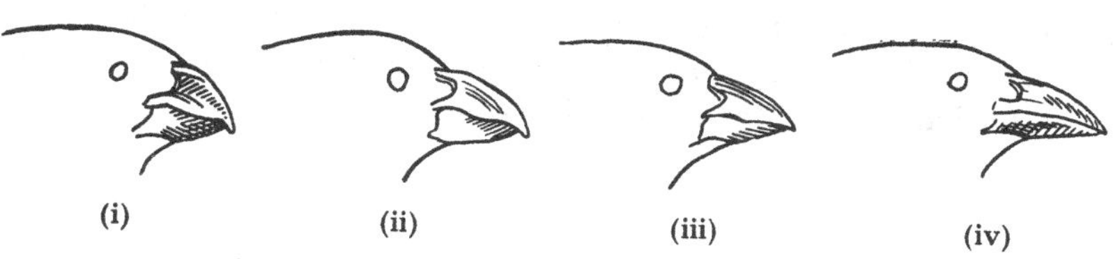

Fig. 13. Beak adaptations in *Camarhynchus*. ⅔ natural size (*after* Swarth).
(i) *C. psittacula psittacula* (James) (ii) *C. psittacula habeli* (Bindloe) (iii) *C. pallidus striatipectus* (Chatham) (iv) *C. pallidus pallidus* (James)

TABLE XIV. BEAK ADAPTATIONS IN *CAMARHYNCHUS*

Species	Island	Comment	Average measurement in mm. Whole culmen	Depth of beak
C. parvulus parvulus	James	*C. psittacula* also present	9·1	7·4
C. parvulus salvini	Chatham	*C. psittacula* absent; beak larger	10·9	7·9
C. psittacula psittacula	James	*C. pallidus* also present	14·1	11·2
C. psittacula habeli	Bindloe	*C. pallidus* absent; beak longer	15·1	10·5
C. pallidus striatipectus	Chatham	*C. psittacula* absent; beak shorter	13·9	8·9
C. pallidus pallidus	James	*C. psittacula* also present	16·7	9·3

Note. The measurements for the whole culmen are taken from Swarth (1931), those for the depth of beak are my own. For numbers measured see Supplementary Table C, p. 184.

CONCLUSION

On the central Galapagos islands there are six subgenera of Darwin's finches, and five of the six show marked beak differences from each other, which are correlated with marked differences in feeding methods. Two of these subgenera, namely

Geospiza and *Camarhynchus*, are further divided into species which differ primarily in size of beak, and these I consider to be adapted for taking partly different foods, though of the same general nature. Some of the finches are absent from outlying Galapagos islands; their food niches may then be filled by different species, or one form may take foods which on the central islands are divided between two species; in both cases there are corresponding beak modifications. To conclude, in Darwin's finches all the main beak differences between the species may be regarded as adaptations to differences in diet.

Chapter VII: SIZE DIFFERENCES BETWEEN ISLAND FORMS

I have not as yet noticed by far the most remarkable feature in the natural history of this archipelago; it is, that the different islands to a considerable extent are inhabited by a different set of beings.... I never dreamed that islands about fifty or sixty miles apart, and most of them in sight of each other, formed of precisely the same rocks, placed under a quite similar climate, rising to a nearly equal height, would have been differently tenanted.... It is the circumstance, that several of the islands possess their own species of the tortoise, mocking-thrush, finches, and numerous plants, these species having the same general habits, occupying analogous situations, and obviously filling the same place in the natural economy of the archipelago, that strikes me with wonder. It may be suspected that some of these representative species, at least in the case of the tortoise and of some of the birds, may hereafter prove to be only well-marked races; but this would be of equally great interest to the philosophical naturalist.

Charles Darwin: *The Voyage of the 'Beagle'*, Ch. xvii

MEASUREMENTS

Several cases of striking beak differences between island forms of the same species were given in the later sections of the previous chapter. Many other island forms show less pronounced beak differences, and some of them also differ in size of body. These differences are for the most part average ones, and are best demonstrated by actual measurements. For this purpose two beak measurements have been taken, one of the culmen (upper edge) from the nostril to the tip of the beak, and the other of the depth of the beak at its base. In addition, the wing was measured from the carpal joint (the main angle) to the tip of the longest primary. In Darwin's finches this latter measurement provides a more reliable indication of general body size than does the total length of the body measured from beak to tail.

In all of Darwin's finches the wing of the male is on the average slightly longer than that of the female, and in all except the warbler-finch *Certhidea* the male has a slightly longer beak than the female. In *Certhidea* the female has a slightly longer beak than the male. These points are shown in Tables XX and XXI (pp. 169, 170). Similar small differences in the average size of the two sexes are found in many other birds, also in many mammals, including mankind. It is not known what advantage accrues to the male in being larger, the female in being smaller, but this relation is so widespread in birds that it presumably has some value. Incidentally in many birds of prey the usual relation is reversed, and the female is the larger sex.

As mentioned in Chapter v (see also Table XX), the average wing-length of *Geospiza* and *Camarhynchus* is greatest in fully black males, slightly smaller in partly black males, and smaller still in males showing no black. Table XXI shows that these different types of males show parallel but very small differences in the average size of the beak. It seems a general rule in song-birds that the adult male should have a slightly longer wing than the first-year male, while in the shrike *Lanius ludovicianus* there is a similar small age-difference in the average size of the beak (Miller, 1931).

Because of the above differences, it is necessary to consider the two sexes separately when comparing the different forms of Darwin's finches. In this book, only the measurements of the males have been tabulated. Those of the females show precisely similar racial and specific variations, so that there is no need to include them as well; they are published elsewhere (Lack, 1945). When considering wing-length, it is also necessary to separate black or partly black males from those in plumage of juvenile type, but the differences in beak measurements between these different types of males are so small that they can be neglected.

In considering racial differences in the average size of beak and wing, it should be pointed out that in many of Darwin's finches these characters do not vary altogether independently of each other. In nearly all the species, those individuals with longer beaks tend also to have deeper beaks, the correlation being particularly marked in the ground-finches *Geospiza fortis*, *G. fuliginosa* and *G. conirostris*, but altogether absent in the woodpecker-finch *Camarhynchus pallidus*. A similar correlation is found in the house sparrow *Passer domesticus* (Lack, 1940*a*), but is absent in most, though not all, forms of the North American *Junco* (Miller, 1941). No other birds appear to have been studied in this respect.

In five of Darwin's finches, namely, the ground-finches *Geospiza magnirostris*, *G. fortis*, *G. fuliginosa* and *G. conirostris* and the tree-finch *Camarhynchus crassirostris*, the individuals with longer wings tend also to have larger beaks, though the correlation is not so marked as that between length and depth of beak. Hence in these species, and particularly in the medium ground-finch *Geospiza fortis*, general body-size presumably has some influence on the size of the beak, though other beak variations are independent of body-size. Neither the house sparrow nor most forms of the junco show a correlation between size of beak and length of wing, though this is found in a few forms of the junco. Correlation coefficients for these characters are set out in Table XXII, p. 171.

DIFFERENCES BETWEEN ISLAND FORMS

For reference purposes, the average measurements of beak and wing are set out in Table XXIII (p. 172) for every species of Darwin's finch on every island from which sufficient specimens have been collected. The three following tables (pp. 174–6) give the limits of measurement of the various forms, and standard deviations are given in Tables XXIX and XXX. These tables show how commonly a species differs significantly in average size on different islands. The differences are often small, many individuals overlapping in measurements with those from other islands, but occasionally they are very marked, with little or no overlap between extreme individuals.

In the large ground-finch *Geospiza magnirostris*, the individuals from the northern islands tend to be rather larger than those from the central islands. The medium ground-finch *G. fortis* and the small *G. fuliginosa* show an opposite tendency, the birds on the southern islands tending to be larger than those on the northern islands, especially in wing-length. But these trends are by no means completely regular, some island populations varying contrarily to the main direction. Further, the differences are only in average size, and there is marked overlap in the measurements of extreme individuals from different islands. The variations in the average length of the wing of *G. fortis* are shown in Fig. 14; for further details see Table XXIII (p. 172).

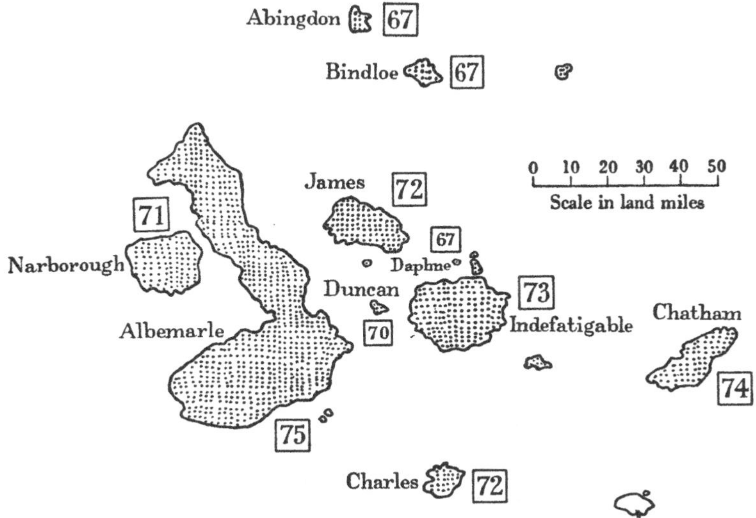

Fig. 14. Variations in wing-length in *Geospiza fortis*. (Average wing-length in mm. for black males.)

The island variations found in the cactus ground-finch *G. scandens* are also erratic, since the small birds from James show no overlap in beak measurements with the large birds from Bindloe, the next island to the north; but on islands to the south occur populations which overlap widely in measurements with both the James and the Bindloe birds. The beak variations in *G. scandens* and the related species *G. conirostris* are shown in Fig. 15. Those of *G. conirostris* were discussed earlier (pp. 66–9).

The marked beak differences between the three subspecies of the sharp-beaked ground-finch *G. difficilis* were also considered in the last chapter, but further variations occur in this species, since each of the three subspecies inhabits two islands, and in each case there are small differences in average measurements between the two island populations (see Table XV, p. 78, and Table XXIII, p. 172).

The vegetarian tree-finch *Camarhynchus crassirostris* and the small insectivorous tree-finch *C. parvulus* show little difference in average size on different islands, except for the unusually large

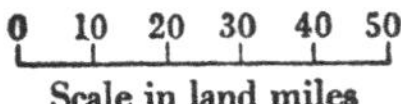

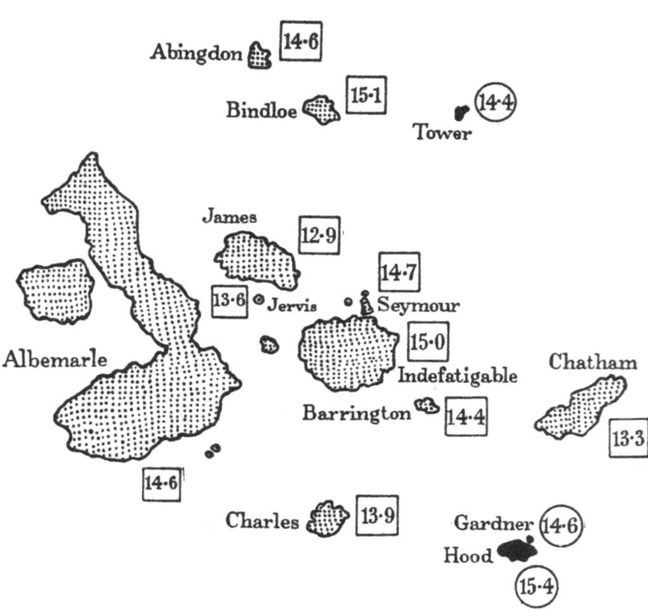

Fig. 15. Variations in beak in *Geospiza scandens* and *G. conirostris*. (Average length of culmen from nostril in mm.)

(i) For *G. scandens*—islands stippled, figures in squares
(ii) For *G. conirostris*—islands black, figures in circles

form of *C. parvulus* on Chatham. On the other hand, both the large insectivorous tree-finch *C. psittacula* and the woodpecker-finch *C. pallidus* show marked island differences; in addition to the variations discussed earlier (pp. 70–1), both species are represented by a particularly small form on Albemarle. Finally, in the warbler-finch *Certhidea* the birds of the outlying Galapagos islands tend to be rather larger than those of the central islands, as set out in Fig. 16. In this species the variations in plumage also tend to a concentric distribution, as shown in Fig. 6 (p. 40).

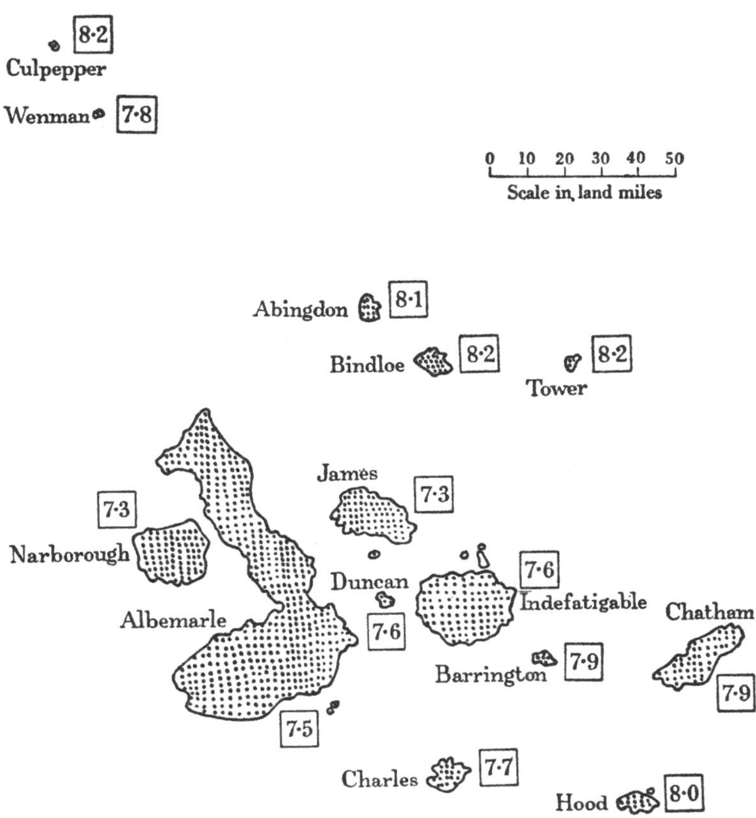

Fig. 16. Variations in beak in *Certhidea*.

(Average length of culmen from nostril in mm.)

NATURE OF THE DIFFERENCES

In Darwin's finches some of the most marked beak differences between island forms of the same species are adaptive, as considered in the last chapter. On the other hand, many of the smaller differences in beak or wing-length are extremely difficult to relate to possible environmental factors. Some examples of this apparently pointless variation are set out in Table XV.

TABLE XV. BEAK AND WING VARIATIONS IN FORMS ON ADJACENT ISLANDS

Island	Culmen (from nostril) in mm. Average	Culmen (from nostril) in mm. Limits	Wing in mm. (black males) Average	Wing in mm. (black males) Limits
(i) *GEOSPIZA MAGNIROSTRIS*—on islands 5 miles apart				
James	15·9	13·8–17·4	84	81–88
Jervis	15·3	13·9–17·0	83	78–90
(ii) *GEOSPIZA SCANDENS*—on islands 5 miles apart				
James	12·9	12·0–13·8	70	66–72
Jervis	13·6	12·4–14·5	71	69–73
(iii) *GEOSPIZA DIFFICILIS SEPTENTRIONALIS*—on islands 20 miles apart				
Culpepper	11·3	10·5–12·2	73	71–77
Wenman	10·7	10·0–11·7	72	68–75
(iv) *GEOSPIZA DIFFICILIS DEBILIROSTRIS*—on islands 10 miles apart				
James	10·3	9·5–11·4	72	65–76
Indefatigable	9·6	9·1–10·2	69	67–71
(v) *GEOSPIZA CONIROSTRIS CONIROSTRIS*—on islands 1 mile apart				
Hood	15·4	13·0–17·4	80	74–84
Gardner nr. Hood	14·6	12·3–17·0	79	69–84

Notes. (i) Similar variations in depth of beak can be found from Tables XXIII, XXIV and XXVI (pp. 172–6).

(ii) The above measurements refer to males only. The females show similar differences in each case.

(iii) The statistical significance of the differences involved can be calculated from the number of specimens measured, and the standard deviations of the measurements, given in Tables XXIX, XXX and Supplementary Table C (pp. 179, 180 and 184).

Table XV shows the existence of significant differences, both in average size and in extreme measurements, between populations of the same species living on islands only a few miles apart. In each case, so far as known, the finch concerned occupies the same ecological niche on both islands, while the two islands provide similar physical and climatic conditions, similar habitats, and the same complement of other species of land birds. In such cases, more of which can be found in Table XXIII, it is almost impossible to believe that the differences between the island forms are adaptively related to possible differences in their environments. The size difference in *Geospiza conirostris* on islands only a mile apart is particularly striking, and recalls the yet more remarkable case of the white-eye *Zosterops rendovae* of the Solomons, in which well-defined island forms are separated by straits which are only 1, 2, 3 and 4 miles wide respectively (Mayr, 1942).

The rarity of regular trends of variation is another fact suggesting that in Darwin's finches many of the differences in beak and wing-length between island forms are without adaptive significance. Even where there is a suggestion of regularity, there are usually certain islands on which variation runs counter to the main trend. Examples of such irregularity can be found in every species.

It is extremely difficult to establish convincingly that a structural difference is non-adaptive. Also, it was a long time before I appreciated that some of the beak differences discussed in the last chapter were adaptive, so that there may well be other differences the adaptive significance of which has not yet been discovered. On the other hand, conditions are extremely similar on many of the Galapagos islands, and the differences in beak and wing-length shown by island forms so often seem haphazard and pointless, that it is probable that many of these differences are genuinely unrelated to possible environmental differences. The ways in which non-adaptive differences might arise between populations are considered in a later chapter.

SIZE VARIATIONS IN OTHER BIRDS

There are many birds in which geographical forms of the same species show average differences in wing-length. For instance, Bergmann's rule describes a general tendency for the races living in cooler climates to have longer wings than those of warmer regions. Again, forest forms tend to be of larger size than related forms in open country, as shown for West African birds by Bates (1931). Similarly, Miller (1931) found that in the shrike *Lanius ludovicianus* the races with the longest wings are those which live in the most open country. Such trends are regular and many forms are similarly affected, so that there is strong evidence for considering that the differences involved are adaptive.

No such regularity occurs in the variations in wing-length found among Darwin's finches, and similar irregular variations occur in many other birds frequenting archipelagos, but are less common in other regions. Thus Swarth (1931) records the average wing-length of the male Galapagos mockingbird *Nesomimus* as varying from 108 mm. on Bindloe to 124 mm. on Hood, and some of the island forms show no overlap with each other even in extreme measurements. Similarly, the male vermilion fly-catcher *Pyrocephalus* has an average wing-length of 57 mm. on Chatham and 64 mm. on Charles, the two populations showing no overlap in the measurements of extreme individuals. These differences are more marked than those found between island forms in Darwin's finches. Instances of similar irregular and marked variations in wing-length are cited by Mayr (1942) for the land birds of other archipelagos. The Polynesian honeyeater *Foulehaio c. carunculata* shows particularly marked and irregular variations (Mayr, 1932*a*). In such cases, as in Darwin's finches, there is usually no reason for thinking that the size differences are adapted to possible differences in the environments of different islands.

Regional variations in the average size of the beak are also fairly frequent in birds. Thus Allen's rule describes a tendency for the forms inhabiting cooler climates to have proportionately smaller beaks and other extremities than have the forms of warmer regions. As this trend is regular, it is presumably adaptive. In continental regions, racial variations in the average size of the beak are usually small, though there are some conspicuous exceptions, as in some races of the European reed bunting *Emberiza schoeniclus*, the nutcracker *Nucifraga caryocatactes* and the crossbill *Loxia curvirostra* (Witherby *et al.*, 1938). In the last species, some of the differences are correlated with differences in diet (Lack, 1944*b*), as discussed further in the next chapter.

Marked beak differences, and marked size differences generally, are commoner in insular than continental races of birds. Where a species is distributed over a big land-mass and also over a number of islands, the races showing marked differences in size, including size of beak, are nearly always those of the islands. Such island forms may be either larger or smaller than the mainland forms, but Murphy (1938*a*) finds that, in the passerine birds of America, the beak is much more often larger than smaller in the insular races. In some cases the beak difference may be due to the island form filling a rather different niche from its mainland relatives, correlated with the paucity of other land birds on the island. In other cases perhaps no adaptive significance is involved. But in nearly all cases too little is yet known of the ecology of the birds for definite conclusions to be reached.

While ecological conditions tend to be different on an island as compared with the mainland, conditions are often closely similar on adjacent islands. Yet some of the most marked beak-variations occur in birds which are distributed over the islands of an archipelago. A particularly good example is provided by the Galapagos mockingbird *Nesomimus*, in which the average beak-length varies from as little as 20 mm. on Albemarle and Indefatigable to as much as 33 mm. on Hood (Swarth, 1931). The long beak of the Hood form is perhaps correlated with its shore-feeding habits, but it is extremely difficult to believe that the beak differences between many of the other forms of the mockingbird are adapted to differences in food or ecology. Similar, apparently non-adaptive, beak differences are found in the island forms of many other land birds on oceanic archipelagos.

CHAPTER VIII: SIZE DIFFERENCES BETWEEN SPECIES

The characters of the species of Geospiza, as well as of the following subgenera, run closely into each other in a most remarkable way.

CHARLES DARWIN: *The Zoology of the Voyage of H.M.S. 'Beagle'*

THE THREE GROUND-FINCHES

THE large ground-finch *Geospiza magnirostris*, the medium *G. fortis*, and the small *G. fuliginosa* differ from each other solely in size and in relative size of beak. The probable connection of this difference with food has been discussed in Chapter VI, but the situation is so remarkable that it is here analysed in further detail.

The limits of measurement given in Fig. 17 and in Table XXIV (p. 174) show that on the northern Galapagos islands the three

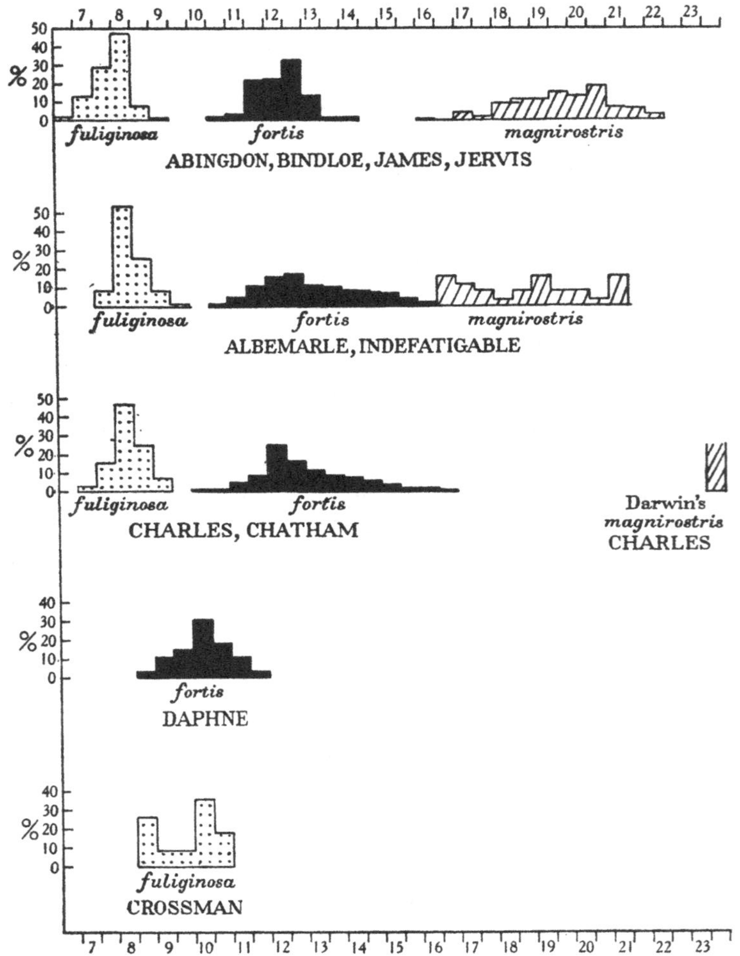

Fig. 17. Histograms of beak-depth in *Geospiza* species.

Measurements in millimetres are placed horizontally, and the percentage of specimens of each size vertically.
Too few specimens are available for a histogram of Darwin's *magnirostris*.
The Daphne and Crossman forms are discussed on pp. 84-86.

species are clearly separated from each other. But on some of the southern islands the separation is very narrow, and occasionally there is actual overlap in characters. Further, on some of the southern islands the medium *G. fortis* is highly variable, and the largest individuals come closer in measurements to small specimens of *G. magnirostris* than they do to small specimens of their own kind, while the latter come closer in measurements to large specimens of *G. fuliginosa* than they do to large individuals of their own kind. This situation is without parallel in birds. The three species, it must be remembered, differ not at all in plumage. It might therefore be wondered whether there are three distinct species, rather than a single highly variable form. But Fig. 17 shows that the measurements of these birds fall into three groups, each centred round a particular mean value, while individuals with intermediate measurements are very scarce.

OVERLAP IN MEASUREMENTS

Table XXIV shows a further peculiarity, namely, that the gap between *G. magnirostris* and *G. fortis*, and the gap between *G. fortis* and *G. fuliginosa*, occur at rather different measurements on different islands. Thus black males with a wing-length of 64–66 mm. belong to *G. fortis* if collected on Abingdon or Bindloe, but to *G. fuliginosa* if from any other island. Similarly, males with a wing-length of 79–80 mm., a culmen from nostril of 14 mm. and a beak depth around 16·0–16·5 mm. belong to *G. magnirostris* if collected on Abingdon, Bindloe or Jervis, but to *G. fortis* if collected on Charles.

As discussed in Chapter VI, the size differences between the species are probably correlated with their taking rather different foods. It is primarily on Charles and Chatham, where *G. magnirostris* is absent, that *G. fortis* attains a large size. However, *G. fortis* also overlaps in measurements with *G. magnirostris* on two islands where the latter species occurs, namely, Albemarle and Indefatigable—but though *G. magnirostris* is present there it is very scarce. Thus of 511 specimens of the two species collected on Albemarle and Indefatigable, only 9 per cent belong to *G. magnirostris*, whereas of 455 specimens from Abingdon, Bindloe, James and Jervis, 55 per cent belong to *G. magnirostris*. *G. fortis* does not normally attain a large size on any of the islands where *G. magnirostris* is common.

Though now absent there, *G. magnirostris* probably occurred formerly on Charles (see pp. 22–3). But the specimens which

Darwin collected are extremely large, and are separated from the largest specimens of *G. fortis* on Charles by a gap as wide as that between the two species on the northern Galapagos islands, where both are smaller. This would suggest that the large individuals of *G. fortis* on Charles did not compete for food with the unusually large Charles form of *G. magnirostris*.

Except on Chatham, where there is a small overlap in their measurements, the gap between the medium *G. fortis* and the small *G. fuliginosa* is more definite than that between *G. fortis* and *G. magnirostris*. But the position is complicated by the

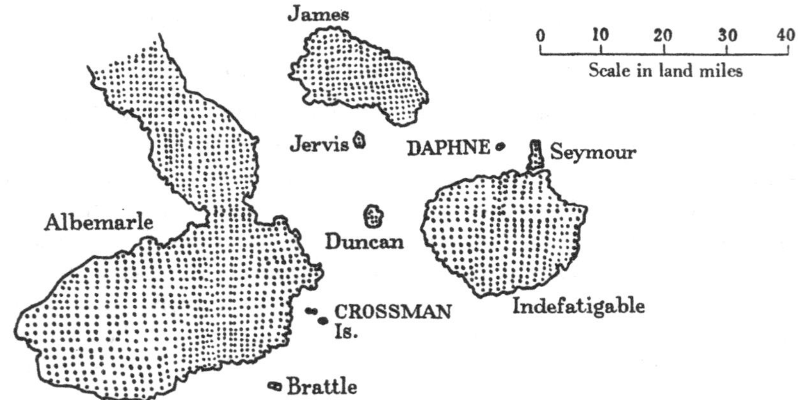

Fig. 18. Map showing Daphne and the Crossman Islets.

existence on the islet of Daphne of an intermediate form, the smallest individuals of which are equal in size to large individuals of *G. fuliginosa* and the largest to small specimens of *G. fortis*. A similar but rather smaller intermediate form occurs on the Crossman islets. As shown in Fig. 18, Daphne is not particularly near the Crossmans, which suggests that these intermediate forms are of independent origin. Formerly, they were referred to a distinct species *G. harterti*, but Swarth (1931) considered that the Daphne birds were unusually small specimens of *G. fortis* and the Crossman birds unusually large specimens of *G. fuliginosa*. As the plumage of the two species is identical, measurements provide the only clue to their specific identity. These are given in Fig. 17 and Table XVI.

TABLE XVI. THE PECULIAR FORMS OF *GEOSPIZA* ON DAPHNE AND CROSSMAN

Form	Culmen from nostril in mm.		Wing in mm. (black males)		Ratio of culmen to wing	Coefficient of variability	
	Mean	Limits	Mean	Limits		Culmen	Wing
G. fortis (Indefatigable)	12·0	(10·5–13·9)	73	(69–79)	0·16	6·8	3·8
Daphne form	10·5	(9·2–11·3)	67	(65–70)	0·16	5·1	2·6
Crossman form	9·3	(8·0–11·2)	66	(63–67)	0·14	9·4	2·3
G. fuliginosa (Indefatigable)	8·4	(7·5–9·3)	64	(61–66)	0·13	4·5	2·2

Note. For number of specimens measured see Supplementary Table C, p. 184.

Formerly I thought that the Daphne and Crossman forms might be of hybrid origin between *G. fortis* and *G. fuliginosa*. However, it would be highly remarkable if two species which occur together on many islands without interbreeding should, just in two places, give rise to a hybrid population. Further, it would be necessary to suppose that such interbreeding no longer takes place, since the Daphne and Crossman birds do not include any individuals in the upper range of size of *G. fortis* nor any in the lower range of size of *G. fuliginosa*. Moreover, the chief character in which these birds are intermediate is the beak, a structure which is particularly adaptable in Darwin's finches.

An alternative explanation is possible. Daphne is only half a mile in diameter and the Crossmans are rather smaller, so that they could support only very small populations of *G. fortis* or *G. fuliginosa*. Perhaps one of the two species became temporarily extinct there, or alternatively the islands may be too small to support populations of both species. If only one of the two species persisted, it might be expected to evolve a beak of intermediate type, since the foods normally taken by both species would be available for it. It has already been shown in Chapter VI that on some of the small outlying Galapagos islands two of the usual ground-finch species are absent, in which case a form has sometimes been evolved which occupies two food niches and has a beak of intermediate type. If a similar explanation holds for the Daphne and Crossman birds, then, to judge from their appearance, the Daphne birds are an unusually small form of *G. fortis* and the Crossman birds an unusually large form of *G. fuliginosa*, as previously concluded by Swarth. This view is corroborated by the ratio of beak-length to wing-length in these forms, as shown in Table XVI.

Ground-finches occasionally wander out to sea (see p. 21), and since Daphne and Crossman are close to large islands, it is probable that typical individuals of *G. fortis* and *G. fuliginosa* occasionally wander there. Indeed, three typical specimens of *G. fuliginosa* have been taken on Daphne. Nevertheless, the peculiar local forms persist, which presumably means that they are better adapted to existence on small islets than are the typical forms of the species concerned. A similar conclusion was reached regarding the persistence of forms such as *G. conirostris* and *G. difficilis septentrionalis*, which combine two food niches on the small remote Galapagos islands (see p. 70).

In addition to the cases of overlap in measurements so far considered, there are four specimens from James and Bindloe which are so intermediate between *G. magnirostris* and *G. fortis* that they cannot certainly be identified. Their measurements are given in Table XXXII (p. 182), but are omitted from Fig. 17 owing to doubt as to the species concerned. In the same table are given the measurements of two specimens, from Hood and Chatham, which are intermediate between *G. fortis* and *G. fuliginosa*; these are probably freak specimens of the smaller species. Such intermediate specimens, which are extremely rare, are discussed further in Chapter x.

DIFFERENCES IN PROPORTIONS

The three species *G. magnirostris*, *G. fortis* and *G. fuliginosa* differ from each other not only in absolute size of beak and wing, but also in their proportions. The large *G. magnirostris* has proportionately the longest beak in relation to wing-length, and proportionately the deepest beak in relation to beak-length. The small *G. fuliginosa* has proportionately the smallest beak in relation to wing-length and proportionately the narrowest beak in relation to beak-length.

Huxley (1927, 1942) has shown in a number of animals that the larger individuals have proportionately larger external parts. This allometric relation holds in such diverse instances as antlers of deer and claws of crabs, and suggests the possibility that the differences in proportions between the three ground-finch species may be simply an indirect effect of their absolute size. To test this, Tables XXVII and XXVIII (pp. 177–8) were prepared. These rule out the above suggestion, showing that within each species the larger individuals do not have relatively larger beaks, while those with longer beaks do not have relatively deeper beaks. Instead, the proportions of wing-length to beak-length, and of length to depth of beak, are approximately the same for all the individuals of each species, whatever their absolute size. Hence the differences between the three species do not depend simply on general size factors with associated allometric relations affecting their proportions.

Since the proportions of beak and wing are characteristic for each of the three species, they provide an additional criterion for identification, but, owing to the marked individual variation, only the average for a series of specimens is reliable. Because of their high variability in both proportions and absolute size, and owing to the narrowness of the gaps between their extreme measurements, and because their plumage is identical, these three ground-finch species show no completely certain criteria for identification. However, nearly every specimen can be safely identified by considering all its characters together, and knowing also the island where it was collected. Presumably the three species have markedly different hereditary constitutions, and so far as known they do not interbreed with each other, but they regularly give rise to individuals which are extremely similar to each other in all external characters, while very occasional individuals are so intermediate in appearance that they cannot safely be identified—a truly remarkable state of affairs. In no other birds are the differences between species so ill-defined.

The photographs shown in Plate VII are of specimens in the California Academy of Sciences. They are all reproduced on the same scale, and show the marked difference in size between the largest and smallest specimen of *G. fortis*, and how closely the largest approaches to *G. magnirostris* and the smallest to *G. fuliginosa*. They also show the difference in the proportions of the beak in the three species.

CAMARHYNCHUS

The insectivorous tree-finches present a situation rather similar to that of the three ground-finches, though with certain differences. On most Galapagos islands there are two species, the large *Camarhynchus psittacula* and the small *C. parvulus*, which have similar plumage and differ solely in size and in proportionate size of beak. As in *Geospiza*, the larger species has proportionately the longer and deeper beak. On Charles there also occurs a third species *Camarhynchus pauper*, intermediate in size between the other two. The size variations in these three species are shown in Fig. 19 and Table XXV (p. 175).

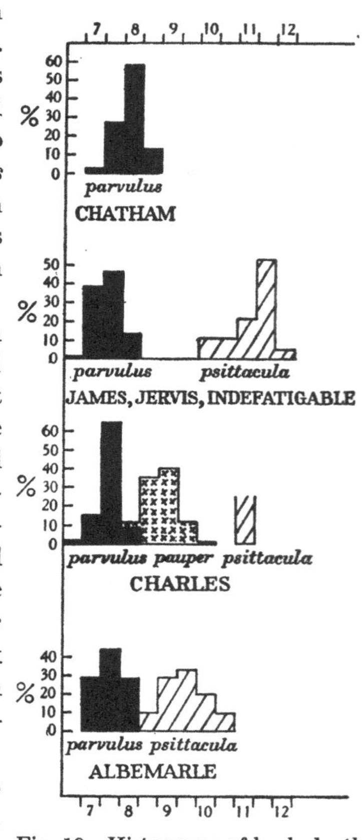

Fig. 19. Histograms of beak-depth in *Camarhynchus* species.

Measurements in millimetres are placed horizontally, and the percentage of specimens of each size vertically.

Too few specimens are available for a histogram of *C. psittacula* on Charles.

The small insectivorous tree-finch *C. parvulus* is of similar size on all the main Galapagos islands except Chatham. Its unusually large size on the latter is probably correlated with the absence there of *C. psittacula* (see p. 63). The largest individuals from Chatham overlap in all measurements with the Albemarle form of *C. psittacula*, but *C. parvulus* does not normally overlap in beak measurements with *C. psittacula* on any island where both occur together.

On the central Galapagos islands of James, Jervis and Indefatigable, the gap between the two species is clear and wide. On Albemarle it is still present, but it is much narrower because of the small size of Albemarle *C. psittacula*. Finally, on Charles the largest specimens of *C. parvulus* are only just separated in depth of beak from the smallest specimens of the medium-sized species *C. pauper*, and in length of wing and culmen from nostril they show a slight overlap. Similarly, the largest specimens of *C. pauper* overlap slightly in length of wing and culmen with the smallest specimens of *C. psittacula*, but they show no overlap in regard to depth of beak, though the gap is extremely narrow. Only sixteen adult specimens of *C. psittacula* have been collected on Charles, as compared with 142 of *C. pauper* and 124 of *C. parvulus*. Evidently the large *C. psittacula* is much less successful than the medium-sized *C. pauper* on this island.

In *Camarhynchus*, as in *Geospiza*, further overlap in measurements is provided by a few freak specimens which are intermediate in size between the small and the larger species. They cannot certainly be identified, but are probably abnormal specimens of *Camarhynchus parvulus*. Their measurements are given in Table XXXII (p. 182).

CASES IN OTHER BIRDS

Examples have already been discussed in Chapter VI of two and occasionally three related species of birds which differ from each other primarily in size and size of beak and share the same habitat. Nearly all such species also differ from each other to some extent in plumage, and the absence of plumage differences, as in *Geospiza*, is unusual. In these other birds the size differences are more clear-cut than they are in *Geospiza* and *Camarhynchus*. But in few other cases are two or three such species found together on the islands of an archipelago, with the favourable conditions thus afforded for the appearance of size differences between local populations; and the position is further complicated in *Geospiza* by the unusually large amount of purely individual variation.

The nearest parallel to *Geospiza* occurs in the crossbills. As mentioned in Chapter VI, there are three species, the large-beaked *Loxia pytyopsittacus* which feeds primarily on pine cones, the medium *L. curvirostra* which feeds primarily on spruce, and the small-beaked *L. leucoptera* which feeds primarily on larch cones. But this is not all, for some of the geographical forms of *L. curvirostra* differ markedly from the typical race in both diet and beak. Thus the Scottish form *L. c. scotica* feeds on pine cones and has a beak closely similar to that of *L. pytyopsittacus*. At the other extreme, the Himalayan form *L. c. himalayensis* has a beak as small as that of *L. leucoptera*, and like the latter species feeds primarily on larch (Lack, 1944*b*). In the crossbills, the correlation between beak and diet is closer than it appears to be in *Geospiza*. But, as already mentioned, the diets of the *Geospiza* species need more detailed study, particularly in the latter part of the dry season, which is likely to be the most critical time.

Chapter IX: INDIVIDUAL VARIATION

No one supposes that all the individuals of the same species are cast in the very same mould. These individual differences are highly important to us, as they afford materials for natural selection to accumulate....I am convinced that the most experienced naturalist would be surprised at the number of cases of variability, even in important parts of structure, which he could collect....It should be remembered that systematists are far from pleased at finding variability in important characters.

Charles Darwin: *The Origin of Species*, Ch. ii

CHARACTERS AFFECTED

Darwin's finches show an unusually large amount of purely individual variation, and this affects all the characters which help to distinguish the various races, species and genera. The striking individual variations in male plumage, both as regards its extent and the age at which it appears, have been considered in Chapter v, while the less marked variations in female plumage were discussed in Chapter iv. The most variable feature of all is probably the beak, thus causing confusion among taxonomists, since this is usually a very stable character in birds. Wing-length is also unusually variable in some of Darwin's finches.

The unusually marked variability in beak is demonstrated most convincingly by the fact that in several cases one island form was earlier thought to consist of two separate species. This happened in the case of the medium ground-finch *Geospiza fortis* on a number of islands, in the small ground-finch *G. fuliginosa* on Chatham, and in the large cactus ground-finch *G. conirostris* on both Hood and Culpepper. The difficulty which these forms present can be gathered from the photographs of extreme individuals of *G. fortis* on Charles, *G. fuliginosa* on Chatham and *G. c. conirostris* on Hood, which are shown in Plate VII. Darwin's finches are also unusually variable in that occasional freak specimens are produced which have departed so far from the normal, particularly in beak, that their species cannot certainly be determined. These freaks are discussed further in Chapter x.

DIFFERENCES BETWEEN ISLAND FORMS

In Darwin's finches the amount of individual variation is sometimes different in populations of the same species on different islands. R. A. Fisher (1937) and others consider on theoretical grounds that small populations are less variable than large ones, and Fisher has demonstrated that this relation holds for variability in birds' eggs. The same hypothesis can be tested in Darwin's finches.

Variability in plumage is difficult to measure statistically. But the measurements of wing-length and size of beak can be utilized, and their standard deviations afford a reliable means of comparing the variability of different forms, provided that the forms are of similar average size. The latter condition holds for island populations of the same species. For the method of calculating the standard deviation, the reader is referred to any text-book on statistics; it is a measure of the scatter of individual measurements round their mean value.

It is difficult to measure accurately the size of the populations of Darwin's finches, but all that is required here is a comparative estimate of abundance. For this purpose, it has been assumed that within any one species the size of each island population is roughly proportional to the area of the island where it is found. The reader can therefore form a rough idea of the size of each population by inspecting the map of the Galapagos (see also pp. 123–4). However, if islands are so close together that birds fly regularly from one to the other, the two populations should be reckoned in together.

The results set out in Table XXIX (p. 179) show that, in the large ground-finch *G. magnirostris*, the populations on the smallest islands, Wenman and Tower, are the least variable, while those on the moderately small islands of Abingdon and Bindloe are somewhat more variable and that on the large island of James is more variable still. But the differences in variability are not so great as might, perhaps, have been expected in view of the marked differences in the size of the islands concerned. Further, the same correlation does not hold on Jervis, since this is a very small island, but its *G. magnirostris* population is rather variable. Since Jervis is only 5 miles from James, it is possible that birds from Jervis mingle to some extent with those from James, though they keep sufficiently separate for the average size of the beak to be significantly different on the two islands. Another difficulty is provided by the population on Indefatigable. This is one of the most variable populations, but *G. magnirostris* is scarce on Indefatigable so that, though the island is large, it is doubtful whether the *G. magnirostris* population is large.

Passing to the medium ground-finch *G. fortis*, the populations on the small islands of Abingdon and Bindloe are less variable than those of the large islands. *G. fortis* also shows a rather small degree of variability on James, which is a large island, and probably also on Jervis, which though a small island is close to James. It seems probable that the reduced variability of *G. fortis* on Abingdon, Bindloe, James and Jervis is due primarily to the fact that the large ground-finch *G. magnirostris* is common there, and that this prevents *G. fortis* from attaining the large size which is possible for it on islands where *G. magnirostris* is absent or very scarce (see pp. 63, 83). Any influence of population size on variability would seem secondary to this ecological factor.

Table XXIX also shows that, in the small ground-finch *G. fuliginosa* and in the warbler-finch *Certhidea*, there is no apparent difference in the variability of populations on small as compared with large islands.

The view that the more abundant forms are the more variable therefore receives but meagre support in Darwin's finches. But the attempt to correlate variability with abundance is justified for structures such as beak and wing only provided that the ecological conditions are the same for the forms concerned. This is not always the case in the Galapagos finches. Thus on islands where the large *Geospiza magnirostris* and the medium *G. fortis* are both common, it might be expected that they would tend to have more restricted diets than on islands where one or other is absent. The presence or absence of *G. fortis* may similarly affect the diet of the small *G. fuliginosa*. Again, in the warbler-finch *Certhidea*, the Hood form feeds on the shore to a greater extent than the other forms. Such ecological differences probably have a greater influence on variability than have differences in the size of the population, so that any possible effects of the latter are much obscured.

The other species of Darwin's finches cannot be utilized in this investigation. Some of them do not breed on the small islands, in others too few specimens have been collected for significant results, and in others the island forms differ too much for direct comparisons to be justifiable. The last particularly applies to those cases in which the bird occupies a different ecological niche on different islands. For reference purposes, the standard deviations of the measurements of these other species are given in Table XXX, p. 180.

There appears to be only one other bird in which the variability of different geographical forms has been compared, namely the North American junco (Miller, 1941). In this bird, the smallest isolated populations are those of *Junco insularis*, *J. bairdi*, *J. vulcani* and *J. aikeni*. When these forms are compared with seventeen more widely distributed forms, they do not show a significant reduction in variability in regard to any of seven measurements affecting wing, tail, leg and beak. It is, of course, possible that the various forms of junco are not exposed to exactly the same ecological conditions.

DIFFERENCES BETWEEN SPECIES

Some species of Darwin's finches are far more variable than others. Precise comparison is difficult because the unit for measuring variability, the standard deviation, is partly dependent on the size of the part measured; species of larger size tend for this reason alone to have larger standard deviations. To compensate for this, it is usual to divide the standard deviation by the mean size, thus obtaining the coefficient of variability, but Day and Fisher (1937) criticize this procedure because it should not be assumed that variability is directly proportional to the mean. However, since it is of interest to compare the variability of different species Table XXXI (p. 181) has been prepared, and the coefficient of variability is used in default of a better unit.

It has sometimes been stated that all of Darwin's finches are exceptionally variable in beak. This is true as regards the occasional appearance of highly freak individuals, which are discussed in Chapter x. But it is not true for most species when variability is measured statistically, by the spread of the normal run of variations round their mean value. The only other song-birds available for comparison are the house sparrow *Passer domesticus*, the shrike *Lanius ludovicianus* and the species of *Junco*. Table XXXI shows that, by the standard of these birds, most of Darwin's finches do not show an unusual degree of spread in their measurements. Unusually marked variability is, however, shown by four forms, the ground-finches *Geospiza magnirostris*, *G. fortis* and *G. conirostris* on all the islands where they occur, and the small ground-finch *G. fuliginosa* on Chatham. Not only do these forms have unusually high coefficients of variability but, as already mentioned, each of the three latter was formerly considered to consist of two distinct species, instead of one. Each is really a single form with no trace of bimodality in characters, though the beak differences between extreme individuals are similar to those found between members of different species or even genera in typical passerine birds.

It would be of great interest to determine the critical factors controlling the variability of each species, and to know why some species are so much more variable than others. Table XXXI shows that, when different species are compared, there is no obvious relation between variability and abundance. It seems probable that in wild birds the most important factors affecting the variability of the beak are first the degree of specialization in feeding habits, and secondly the extent of competition for food with other species, these two factors being partially interrelated. It seems significant that all the highly variable forms of Darwin's finches are generalized finch-like types with unspecialized diets. On the other hand, some of the least variable species have specialized beaks or feeding habits, such as the woodpecker-finch *Camarhynchus pallidus*, the vegetarian tree-finch *C. crassirostris*, and the Cocos-finch *Pinaroloxias*. However, another bird with specialized feeding habits, namely, the cactus ground-finch *Geospiza scandens*, is one of the more variable forms.

In the Galapagos there are only a very small number of

passerine species to share out the available foods. It might therefore have been expected that Darwin's finches would have more varied diets, and hence more variable beaks, than continental passerine birds. However, to set against this, there is also a much smaller variety of land plants and insects in the Galapagos than in a continental area. In fact, most of Darwin's finches show a degree of variability in beak about equal to that of continental passerine birds. Why this should be so, and why, at the same time, just a few forms should be exceptionally variable, only further research can determine.

Chapter X: HYBRIDIZATION

The view generally entertained by naturalists is that species, when intercrossed, have been specially endowed with the quality of sterility, in order to prevent the confusion of all organic forms.... I hope, however, to show that sterility is not a specially acquired or endowed quality, but is incidental on other acquired differences.

Charles Darwin: *The Origin of Species*, Ch. VIII

FIELD OBSERVATIONS

In certain groups of plants there occur peculiar 'species-swarms', many closely related forms being found in the same region and differing from each other in comparatively few characters. These swarms are partly the result of hybridization between species, with resulting recombination of their characters, and their occurrence has led Lowe (1930, 1936) to suggest the same manner of origin for the not dissimilar 'swarm' of Darwin's finches. Certain other writers have also suggested that there is an unusually large amount of interbreeding between the species of Darwin's finches.

With the above point in mind, a careful watch was kept throughout the breeding season of 1938–9 for possible cases of interbreeding. None was observed, nor have any been recorded by Beck, Gifford or the other experienced collectors who have worked in the Galapagos. Clearly hybridization between species is rare, if not absent. This is corroborated by the behaviour of the captive ground-finches brought to California. Among these birds, individuals of the same species have bred freely with each other, but it has not so far proved possible to induce individuals of different species to breed together. The species involved in these experiments were the ground-finches *Geospiza magnirostris*, *G. fortis*, *G. fuliginosa* and *G. scandens* (Orr, in the Press).

FORMS OF INTERMEDIATE APPEARANCE

The large cactus ground-finch *G. conirostris* on Hood and Tower has a beak intermediate in character between that of the medium ground-finch *G. fortis* and that of the cactus ground-finch *G. scandens*. It might therefore be suggested that *G. conirostris* is of hybrid origin between the two latter species, but there is no evidence for this, and the intermediate nature of its beak is much more satisfactorily explained by supposing that *G. conirostris* combines the feeding rôles of the other two species, as discussed on pp. 66–9. Similarly, the sharp-beaked ground-finch *G. difficilis septentrionalis* of Wenman and Culpepper has a beak intermediate between that of typical *G. difficilis* and that of the cactus ground-finch *G. scandens*. This also is to be regarded as an adaptive modification to a particular food niche, and there is no reason to think that the bird is of hybrid origin (see pp. 67–8). Again, the medium ground-finch *G. fortis* is intermediate in every way between the large *G. magnirostris* and the small *G. fuliginosa*, but there is no reason whatever for considering that it is a hybrid between them. These cases show that in Darwin's finches a beak of intermediate character does not usually imply a hybrid origin, but rather that the form concerned occupies an intermediate ecological niche.

More puzzling is the large-beaked ground-finch on remote Culpepper. Originally Rothschild and Hartert (1899, 1902) named this bird *G. darwini*, but later they classified it as a new race of the large cactus ground-finch *G. conirostris*. Some of the later specimens were referred by Swarth (1931) to the large ground-finch *G. magnirostris*, others to the Tower form of *G. conirostris*; but they all really belong to *darwini*. The beak of *darwini* shows clear basic affinities with that of other forms of *G. conirostris*, but is heavier, with superficial similarities to that of *G. magnirostris*. It is also extremely variable. These facts originally led me to suggest that *darwini* was of hybrid origin between *G. conirostris* and *G. magnirostris*. Pointing against this, on Tower these two species reside together without interbreeding. Further, *darwini* possesses characters of female plumage shown by neither *G. magnirostris* nor *G. conirostris* elsewhere, namely, olive tips to the back feathers, buff on the underparts, and rufous on the wing. These appear to be primitive plumage characters, as discussed in the next chapter (pp. 101–2). Probably *darwini* is a genuine form of *G. conirostris* with a beak modified for ground-feeding (see p. 68), but the possibility of a hybrid origin cannot be altogether excluded. Its exceptional variability in beak is difficult to explain, but the bird possibly has an unusually varied diet. The other forms of *G. conirostris* are also highly variable in beak, though not to the same extent.

The peculiar forms of *Geospiza* on Daphne and Crossman might also be thought to be of hybrid origin, since they are intermediate in appearance between the medium *G. fortis* and the small *G. fuliginosa*. But, as already discussed (pp. 84–6), their intermediate appearance is more probably due to occupation of an intermediate ecological niche.

The small ground-finch *G. fuliginosa* is unusually variable on Chatham, where both very small and very large individuals occur (see Tables XXIV and XXIX, pp. 174, 179). This suggests the possibility that it is a racial hybrid between a small form of *G. fuliginosa* from the northern islands and a larger form from the southern islands, but so little is yet known of the factors controlling variability that this explanation cannot be considered at all certain, and there may well be some other reason.

FREAK SPECIMENS

In Darwin's finches there are no other species or island forms for which a hybrid origin might be suspected. But the collections include a number of odd specimens which are difficult or impossible to assign to known species, and which in many cases are intermediate in appearance between two species. Some of these specimens were formerly thought to be new species, and were named as such, but they are certainly too rare for this, and they must be regarded as either freaks or hybrids. The specimens concerned are listed, together with their wing and beak measurements, in Table XXXII, p. 182.

The dwarf specimen of the vegetarian tree-finch *Camarhynchus crassirostris* can confidently be identified as a freak and not as a hybrid with some other species. On the other hand, the remarkable '*Camarhynchus conjunctus*' and '*Camarhynchus aureus*' are probably hybrids between the warbler-finch *Certhidea* and the small insectivorous tree-finch *Camarhynchus parvulus*, dissimilar though these two genera are. As shown in Fig. 20, these peculiar forms have a beak intermediate in appearance between that of *Camarhynchus* and *Certhidea*, and in addition *C. conjunctus* shows the chestnut throat-patch typical of *Certhidea*. The peculiar '*Cactospiza giffordi*' also has a chestnut throat-patch, but in other respects looks like a dwarf woodpecker-finch *C. pallidus*, and is either a freak specimen of the latter or possibly a hybrid between this species and *Certhidea*.

Fig. 20. Possible intergeneric hybrids.
$\frac{2}{3}$ natural size (*after* Swarth).

(i) *Camarhynchus parvulus parvulus* (ii) '*Camarhynchus aureus*' (Chatham)
(iii) '*Camarhynchus conjunctus*' (Charles) (iv) *Certhidea olivacea* (Charles form)

The nature of the other specimens is much more problematical. Related species of Darwin's finches look very similar, and in some cases differ solely in total size and relative size of beak. An unusually small specimen of a large species, an unusually large specimen of a small species, and a hybrid between them, might well be indistinguishable in appearance. This difficulty has already become apparent when discussing the affinities of the ground-finches of Daphne and Crossman, and it also applies to many of the specimens listed in Table XXXII. Thus the peculiar specimens intermediate between *Geospiza fortis* and *G. scandens* might be freaks of either species or hybrids between them; their resemblance in beak to another species, *G. conirostris*, is presumably fortuitous and further illustrates the difficulties in identifying such individuals. It is unfortunate that the captive finches brought to California have so far refused to interbreed, as specimens of known hybrid origin might have helped to clear up the relationships of the freak specimens taken in the field. As shown in Table XXXII, such abnormal or intermediate specimens occur in nearly all the species of Darwin's finches.

HYBRIDIZATION AND THE ORIGIN OF NEW FORMS

Unusually variable island forms, attributable to hybridization between races of the same species, have been described by Mayr (1932*b*, 1938) for the flycatcher *Pachycephala* in the Fijis and the Solomons, and also for the mound-builder *Megapodius eremita* on Dampier Island off New Guinea, while in continental birds adjacent subspecies frequently interbreed in a zone of contact. In some of these cases the hybrid population is highly variable, and ranges in appearance between both parental types, while in other cases, such as that of *Junco hiemalis cismontanus* (Miller, 1941), a more stable form is produced. In cases of the former type, variability is perhaps maintained by repeated fresh contacts between the parent races, while in the stable forms some of the hereditary factors of the parent races have evidently been eliminated.

Hybrids between species are much rarer than interracial hybrids, and hybrids between members of different genera are rarer still, particularly among birds living in a wild state. However, natural intergeneric hybrids have been recorded between the greenfinch *Chloris chloris* and the goldfinch *Carduelis carduelis*, and between the swallow *Hirundo rustica* and the house martin *Delichon urbica* (Meise, 1936*a*; Mayr, 1942), so that parallels exist with the probable hybrids between *Certhidea* and *Camarhynchus parvulus* in the Galapagos.

The rarity of hybrid birds suggests that they are less efficient than either parent species. Frequently they are sterile, and Dobzhansky (1937) considers that they also tend to be less well adapted in other ways. Their disadvantages tend to be greater the greater the degree of difference between the two parental types, and so are greater in a hybrid between two species than in one between two races of the same species. Occasionally in birds a form of hybrid origin between two races has become established, showing that these disadvantages may be overcome, but there is only one known case of an established form of hybrid origin between two species. This is the hybrid between the house and Spanish sparrows *Passer domesticus* and *P. hispaniolensis*, which is established to the exclusion of both parent species in certain North African oases, and perhaps also in Italy (Meise, 1936*b*).

The survey by Huxley (1942) shows that, in plants, established forms of hybrid origin are much commoner than they are in birds. But most successful plant hybrids are allopolyploids, which means that they possess the complete set of inherited factors of both parent species. As a result, they do not usually suffer from the disadvantages associated with hybrid forms in most animal groups, and they tend to be fertile. Polyploidy is unknown in birds, so that the comparison between Darwin's finches and a species swarm in plants is not a valid one.

It is therefore improbable on general grounds that hybridization between species has played an important part in the origin of new forms of birds. In Darwin's finches no cases of interbreeding between species have been observed in the wild, and individuals of different species have not as yet been induced to interbreed in captivity. Some forms of Darwin's finches are intermediate in appearance between two species, but in most cases this is probably due to the intermediate nature of their ecological requirements and not to a hybrid origin. There are also a number of freak specimens, but it is not certain how many of these are hybrids, and their rarity indicates that they are at a disadvantage, though selection is evidently less strict than is the case in most birds. To conclude, it seems probable that hybridization has not played an important part in the origin of new forms of Darwin's finches.

CHAPTER XI: AN EVOLUTIONARY TREE

On the principle of the multiplication and gradual divergence in character of the species descended from a common parent, together with their retention by inheritance of some characters in common, we can understand the excessively complex and radiating affinities by which all the members of the same family or higher group are connected together.

CHARLES DARWIN: *The Origin of Species*, Ch. XIII

THE ORIGINAL STOCK

FROM the seeds planted by Darwin, a forest of evolutionary trees came to adorn the text-books of zoology. Their cultivation is now somewhat out of fashion, but since the central theme of this book is that Darwin's finches evolved from a common stock, it is necessary to suggest the steps by which this could have come about, though in this matter no finality of judgment is possible. The difficulties in reconstructing the course of evolution are great, since resemblances between existing species may be due to close relationship, but may also be brought about by parallel evolution, or by the chance retention of the same primitive features. Further, island populations are often small, so that forms can become modified or extinct very quickly. Nor are there any fossil specimens to help fill the gaps in the living record.

Previous chapters have shown so many resemblances between the various species of Darwin's finches that it is scarcely necessary to reaffirm that they are all related to each other. All recent writers are agreed on this, Swarth (1931) from a study of skins, Snodgrass (1903), Sushkin (1925, 1929) and Lowe (1936) on anatomical grounds, and the present writer from a study of their breeding behaviour. But though all the species show marked similarities to each other, they do not show a close resemblance to any particular species of finch on the South or Central American mainland. Either the mainland ancestor has become extinct, or Darwin's finches have diverged from it so far that their close relationship is no longer apparent. Sushkin and Lowe have established that Darwin's finches are derived from the fringilline subfamily of the finches, but it is difficult to restrict their point of origin more closely.

The evidence suggests that the following are primitive geospizine features: black plumage in the male, streaked underparts and a rufous wing-bar in the female, a heavy finch-like beak, a diet of seeds, and a habitat in the arid lowland zone. One of Darwin's finches, namely, the sharp-beaked ground-finch *Geospiza difficilis*, possesses all these characteristics. The differentiation of this species into three well-marked races also suggests that it is a long-established form, while its irregular distribution on the outlying Galapagos islands suggests that it is in process of elimination by the small ground-finch *G. fuliginosa*, which was presumably evolved later (see pp. 26–8).

The Culpepper and Wenman form of *G. difficilis* has an olive tinge to the upper parts, a buff tinge to the underparts and a rufous wing-bar, characters which are shared by two other species, namely, the peculiar Culpepper ground-finch *G. conirostris darwini* and the Cocos-finch *Pinaroloxias*. Such a discontinuous distribution of characters suggests that they are primitive. Hence *Geospiza difficilis septentrionalis* of Culpepper and Wenman can probably be regarded as the least modified living representative of Darwin's finches.

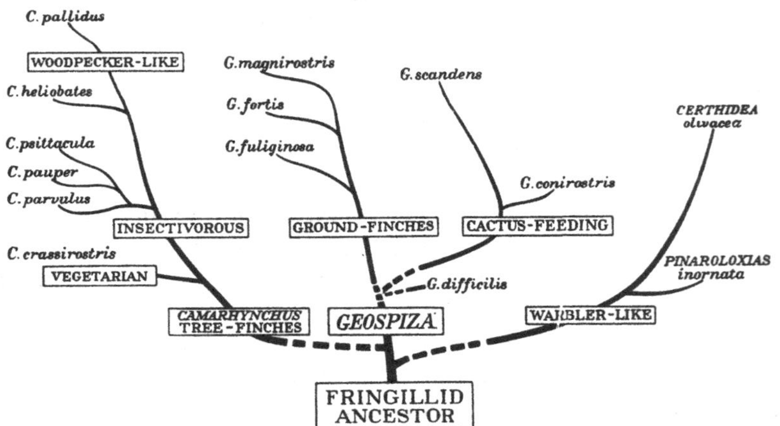

Fig. 21. Suggested evolutionary tree of Darwin's finches.

THE BRANCHES

The small ground-finch *G. fuliginosa* shows many of the features which are presumed to be primitive in Darwin's finches, but it lacks the rufous wing-bar. It may well have evolved from the sharp-beaked ground-finch *G. difficilis*, or from an ancestral form of the latter. Indeed, the Tower form of *G. difficilis* looks so like *G. fuliginosa* that, had *G. difficilis* occurred only on Tower (where *G. fuliginosa* is absent), the two would probably have been regarded as geographical races of the same species. It is even possible that *G. fuliginosa* was derived from *G. difficilis* via the Tower form of the latter, but I would not press this point, as I now consider that some of their resemblances are due to parallel evolution (see p. 67).

The medium ground-finch *G. fortis* differs from the small *G. fuliginosa* solely in size and proportions, and extreme individuals of the two species approach each other closely. Hence *G. fortis* may well have evolved from the smaller species. Similarly, the large ground-finch *G. magnirostris* differs from the medium *G. fortis* solely in size and proportions, and very possibly evolved from it.

The relationship of the large cactus ground-finch *G. conirostris* is more difficult to determine. The discontinuous distribution of this bird solely on the three outlying islands of Culpepper, Tower and Hood, together with its differentiation into well-marked races, indicate that it is a long-established species. This is corroborated by the fact that the Culpepper form *G. c. darwini* possesses primitive plumage features found otherwise only in the Cocos-finch *Pinaroloxias* and in the Culpepper and Wenman form of the sharp-beaked ground-finch *Geospiza difficilis*. Probably *G. conirostris* was derived from an ancestral form of *G. difficilis*, and its present distribution suggests that it may at one time have been much more widespread in the islands than it is to-day. Possibly it occupied the large ground-finch niche on many islands until the later evolution of *G. magnirostris* and *G. fortis*.

The cactus ground-finch *G. scandens* agrees with *G. conirostris* in its dark underparts and its habits of feeding on *Opuntia*. Further, it replaces *Geospiza conirostris* geographically. *G. scandens* is more specialized than *G. conirostris* in the cactus-feeding rôle, and correlated with this it has a longer and thinner beak. But the gap between the two species is bridged by the comparatively narrow-beaked form of *G. conirostris* on Tower and the comparatively broad-beaked forms of *G. scandens* on Abingdon and Bindloe, which are adjacent to Tower. It is very possible that *G. scandens* was derived from *G. conirostris*, and they should perhaps be regarded as well-marked geographical forms of the same original species. Certainly they are closely related. The Culpepper and Wenman form of the sharp-beaked ground-finch *G. difficilis* feeds on cactus, and there is no difficulty in supposing that *G. conirostris* and *G. scandens* evolved from the same stock as *G. difficilis*.

The tree-finches *Camarhynchus* were presumably derived from *Geospiza* stock by partial loss of black plumage in the male, reduced streaking in the female, and some modification in beak and feeding habits. The *Camarhynchus* line early diverged into two, the insectivorous forms on the one hand and the vegetarian tree-finch *C. crassirostris* on the other. The latter species has a distinctive song and a somewhat distinctive beak, but resemblances in plumage and beak indicate relationship with the insectivorous forms.

The insectivorous tree-finches have diverged further, into the small *C. parvulus* and the large *C. psittacula*. These two species differ only in size, and one presumably evolved from the other. The evolution of the pronounced island forms of *C. psittacula*, and the occurrence of two similar species *C. psittacula* and *C. pauper* together on Charles, is discussed later (pp. 126–8). Another insectivorous species, the woodpecker-finch *C. pallidus*, is very similar to the other species in song, plumage, feeding habits and beak, and obviously evolved from the same stock, but it has become more specialized for tree-climbing, wood-boring, and probing into cracks. Finally, the mangrove-finch *C. heliobates* shows considerable resemblance in beak to *C. pallidus* and is presumably closely related to it. Its darker plumage suggests that it may be more primitive than *C. pallidus*; it is not known whether it possesses the latter's specialized feeding habits.

An offshoot in a different direction is the Cocos-finch *Pinaroloxias inornata*. This bird has a slender beak like that of the warbler-finch *Certhidea*, but in plumage it shows marked affinities with the sharp-beaked ground-finch *Geospiza difficilis*, and particularly with the Culpepper and Wenman form, *G. d. septentrionalis*. The males of *Pinaroloxias* and *Geospiza d. septentrionalis* agree in having wholly black plumage save for rufous or buff under-tail coverts, and the females agree in having buff-tinged and much streaked underparts, a rufous wing-bar and an olive tinge to the upper parts. The marked specialization of its beak indicates that *Pinaroloxias* is an early offshoot from the geospizine stock, and the several resemblances in plumage to *Geospiza difficilis* are presumably primitive characters which it has retained.

Finally, the warbler-finch *Certhidea olivacea* is so distinctive that it obviously diverged from the main geospizine stock very early. That it is a long-established form is also suggested by its differentiation into well-marked island forms. It agrees with the other species of Darwin's finches in internal anatomy, in display and nesting habits, and in the possession (by some island forms) of a rufous wing-bar. Its song is more distinctive than that of other species, but shows similarities to that of the sharp-beaked ground-finch *Geospiza difficilis* on Tower. *Certhidea* differs from *Geospiza* in having unstreaked underparts, but juveniles of the Charles form of *Certhidea* are heavily streaked, while streaking has been lost in some species of *Camarhynchus*. *Certhidea* is also peculiar in possessing an orange-tawny throat patch in the male, but this occurs sporadically in other forms (see p. 51). The freak variants '*Camarhynchus conjunctus*' and '*C. aureus*' (see p. 98) are a further link between *Certhidea* and *Camarhynchus*, but they are perhaps hybrids, and their existence does not necessarily imply that *Certhidea* was derived from *C. parvulus*. The most peculiar feature of *Certhidea* is its slender warbler-like beak, but one other species, namely the Cocos-finch *Pinaroloxias*, has a slender beak. This may be due to parallel evolution, but it also seems possible that *Certhidea* and *Pinaroloxias* were derived from a common stock, and that the further modifications in *Certhidea*, particularly in plumage, appeared later.

The foregoing discussion may not be correct in every detail, but its purpose is to show the marked similarities between all of Darwin's finches, and that the derivation of these birds from a common finch-like stock presents no particular difficulties. The precise steps of this evolution cannot be known with certainty.

SUCCESS

Evolutionary moralists have been much concerned with the question of whether specialization pays. At first sight the most successful of Darwin's finches would appear to be the warbler-finch *Certhidea*, as it is the only species to occur on every island, it has a wider habitat range than any other species, and nearly everywhere it is found in greater numbers than any other form. From this, it might be concluded that specialization is advantageous. But there are a number of difficulties. First, though *Certhidea* is the least finch-like of Darwin's finches, and therefore in one sense the most specialized, if regarded as an insectivorous bird it is rather generalized. Further, its present success is probably due to the paucity of true American warblers in the Galapagos, and so is in a sense accidental, and perhaps only temporary. Finally, the comparison of *Certhidea* with any one other species of Darwin's finches may be unfair. If, as is possible, *Certhidea* became split off from the ancestral stock before the genus *Geospiza* had diverged into its present species, then it should more properly be compared with the whole genus *Geospiza*. This leaves the advantage with *Geospiza*, as it is divided into five species, whereas there is only one species of *Certhidea*. Such considerations make further pursuit of this subject unprofitable, and the whole question is somewhat meaningless.

PART TWO: INTERPRETATION

CHAPTER XII: THE ORIGIN OF THE GALAPAGOS FAUNA

All the foregoing remarks on the inhabitants of oceanic islands—namely, the scarcity of kinds—the richness in endemic forms in particular classes or sections of classes—the absence of whole groups, as of batrachians, and of terrestrial mammals notwithstanding the presence of aerial bats—the singular proportions of certain orders of plants—herbaceous forms having been developed into trees, etc.—seem to me to accord better with the view of occasional means of transport having been largely efficient in the long course of time, than with the view of all our oceanic islands having been formerly connected by continuous land with the nearest continent; for on this latter view the migration would probably have been more complete; and if modification be admitted, all the forms of life would have been more equally modified.

CHARLES DARWIN: *The Origin of Species*, Ch. XII

ORIGIN OF THE ISLANDS

THERE has been much speculative dispute as to whether the Galapagos were formed from volcanoes pushed up out of the sea, or alternatively whether they were once connected by land with the American continent. No clue is provided by the rock of which the islands are made, as this is almost entirely volcanic, rather newer in type than that composing the Andes. In addition, there are some raised beaches containing fossil molluscs of pleistocene and tertiary (probably pliocene) age (Dall and Oschner, 1928*a*). The presence of raised beaches shows that there has been some elevation of the land in the Galapagos from the time of the pliocene onward, but it is the period prior to this which is chiefly of interest in the present discussion.

Previous to the publication of *The Origin of Species*, it seems to have been assumed that oceanic islands received their animal and plant life as a result of former land connection with one of the continents. But Darwin (1859), whose argument is summarized at the head of this chapter, concluded that the Galapagos and other oceanic islands arose from the sea, and that their animal and plant life arrived later by occasional means. In this view he was followed by Wallace (1880) and, among recent authors, by Swarth (1934). On the other hand, Baur (1891), Scharff (1912), Van Denburgh and Slevin (1913) and Beebe (1924) considered that the Galapagos animals could have reached the islands only by means of a former land connection with America. The latter view postulates a submergence of the intervening land or an elevation of the sea of some 2000 fathoms, a change of no mean magnitude, the possible mechanics of which are difficult to conceive.

On the American mainland opposite the Galapagos there live several hundred species of passerine and near-passerine birds—the latter term being used to include the woodpeckers, cuckoos and a few other groups which are closely related to the passerine order. The Galapagos provide a variety of habitats suitable for such birds, including mangrove swamp, cactus semi-desert, humid forest and open uplands. Yet the descendants of only six passerine forms and one species of cuckoo live there. Similarly, despite the rich variety of American mammals, reptiles and amphibia, no amphibia, only two types of land mammals and five of land reptiles are established in the islands. The land mammals consist of the rice-rat *Nesoryzomys* and the bat *Lasiurus brachyotis*, and the land reptiles include the giant tortoise *Testudo*, the land and marine iguanas *Conolophus* and *Amblyrhynchus* (which very probably evolved from one original colonist), a snake *Dromicus*, a lizard *Tropidurus* and a gecko *Phyllodactylus* (Van Denburgh, 1912–14). Likewise only an extremely restricted number of land insects and land molluscs are present in the islands (Gulick, 1932). There are also great gaps among the land plants, including a complete absence of conifers, palms, aroids and Liliaceae, while several important families of tropical American dicotyledonous plants are also absent or very poorly represented, such as the Lythraceae, Malastomaceae, Myrtaceae, Onagraceae, and Sapindaceae (Robinson, 1902).

Had there once been a land connection between the Galapagos and America, it seems inconceivable that so few land animals and plants should have availed themselves of it. But this paucity of colonists is just what is to be expected if the only way for land organisms to reach the islands has been by crossing at least 600 miles of uninterrupted ocean. Moreover, the forms which have arrived show an exceedingly varying degree of divergence from their mainland relatives, some being almost or quite indistinguishable and others extremely distinctive. Doubtless the various Galapagos animals have become modified at different rates, but the marked differences in the degree of their divergence strongly suggest that they have been isolated in the Galapagos for very varying periods of time—which is to be expected if they have reached the islands only more or less accidentally by crossing the ocean.

The exponents of the land-bridge hypothesis put forward two main arguments. First, where an oceanic island has been colonized only by occasional means across the sea, many gaps are to be expected in the land flora and the land fauna, which will as a result appear 'disharmonic' when compared with a continental habitat. From this point of view Baur (1891) rather surprisingly claimed that the land life of the Galapagos was harmonic, and therefore that it was part of a continental flora and fauna later cut off by the sea. But the harmonic appearance of the Galapagos land life is only superficial. For instance, the diversity of the land birds is due largely to the adaptive radiation within the one group of Darwin's finches, and most of the ordinary mainland types are absent. Again, while much of the Galapagos is covered with trees, the number of species is few for a tropical environment, and some of them are highly unusual. Members of the family Compositae rarely form trees, but the helianthoid plant *Scalesia* is a dominant tree in the Galapagos humid forest. Likewise, on the American mainland the prickly pear *Opuntia* grows as low scrub, but in the Galapagos it is a tall tree. Herbivorous land mammals are absent from the Galapagos, but their place is filled by giant tortoises and iguanas. There are few land molluscs, but one of these, *Nesiotus*, has produced a diversity of forms (Gulick, 1932). These cases show that the apparent 'harmony' which Baur observed in the Galapagos is a secondary phenomenon. It is true that the chief ecological niches are filled, but many of them are filled by unusual types derived from comparatively few original forms, so that the arguments under this head really support the view that the islands are oceanic in origin.

The only other important argument in favour of a former land bridge is the difficulty of conceiving by what other means the reptiles, and particularly the giant tortoise and the iguanas, could have reached the islands. Van Denburgh and Beebe have particularly stressed this difficulty. However, Simpson (1943) has now shown that the ancestors of the Galapagos tortoise were present on the South American mainland as early as the miocene, which means that they must have arrived there during the early tertiary, when the continent was isolated by sea. In the case both of its colonization of the Galapagos, and its earlier colonization of the South American continent, Simpson considers that the tortoise crossed over the sea, as it can float and survive for long periods in sea water. Another group of giant land tortoises, of independent origin from those in the Galapagos, occurs on the Mascarene Islands in the Indian Ocean, islands which there is no other reason to think were formerly connected with a continent. It therefore seems probable that tortoises can occasionally cross the sea, though there is no direct evidence as to how they do so. The transport of small reptiles over the ocean is not so difficult to imagine, and some authorities consider that the large size of the Galapagos tortoise and iguanas was probably evolved after their arrival in the islands; Brobdingnagians are essentially an insular creation. However, H. W. Parker informs me that this conclusion cannot be considered certain, the fossil remains on the continents showing that there were also giants in the earth in those days.

There is now a fair body of evidence in favour of supposing that small reptiles occasionally cross quite wide stretches of ocean. Thus Dr Mayr informs me that lizards have successfully colonized much of Polynesia, including New Caledonia and many other islands which, from every other point of view, seem typically oceanic and could not have been connected by land bridges. He also cites instances from remote islands in other parts of the world. The way in which small reptiles cross the sea is still unknown, perhaps floating in the water, perhaps on floating vegetation, perhaps during hurricanes. Since reptiles have become established in the Galapagos only five or six times in hundreds of thousands and perhaps millions of years, it is evident that such transport over the sea is extremely rare, so it is scarcely surprising that no biologist has yet witnessed it.

The land reptiles and the rice-rats are found on most of the Galapagos islands. This might suggest that the islands were once connected by land with each other but, though this is a simpler hypothesis than that of a former land bridge with America, it still postulates a change in sea-level of a thousand fathoms. It is therefore simpler to suppose that the reptiles crossed to each Galapagos island over the sea. If these animals could make the 600-mile sea crossing from the American mainland, the 10-, 20- and 30-mile gaps between the individual Galapagos islands should not present an insuperable obstacle. That both the reptiles and the rice-rats are divided into well-marked island forms shows that even such comparatively short sea crossings have been made only rarely.

While almost the whole of the Galapagos land fauna and flora is of American origin, one of the land molluscs has travelled from considerably further. *Tornatellides chathamensis* is the sole Galapagos representative of a group characteristic of Polynesia and absent from America. This species probably came from islands at least 3000 miles to the west of the Galapagos (Gulick, 1932), and since it is the sole Polynesian element in the Galapagos land fauna, a land connection with Polynesia seems quite out of the question.

GALAPAGOS LAND BIRDS

Wallace (1880) drew attention to the very varying degree of divergence shown by the Galapagos land birds, and suggested that this was correlated with their having been a varying length of time in the islands. The cuckoo *Coccyzus melacoryphus* is identical with a South American species, so is probably a recent colonist. The warbler *Dendroica petechia aureola* is considered an endemic race, but it is extremely similar in appearance to the Ecuadorean form of the same species (Chapman, 1926). The martin *Progne m. modesta* is also an endemic Galapagos subspecies of a mainland species. The Galapagos tyrant flycatcher *Myiarchus magnirostris* is sufficiently distinctive to rank as a separate species, but there are closely related species on the American continent. The Galapagos vermilion flycatcher *Pyrocephalus rubinus* is usually classified in the same species as a bird of the American mainland, but in another respect evolution has proceeded further than in the case of *Myiarchus*, since three island races have become differentiated within the Galapagos, one on Chatham, a less distinctive form on Indefatigable, and a third form found on many islands (Hellmayr, 1927; Swarth, 1931).

Evolution has proceeded considerably further in the Galapagos mockingbird, which has obvious affinities with the mainland genus *Mimus*, but is sufficiently distinctive to be placed in a separate genus *Nesomimus*. No island has more than one form, but the forms of this bird on Chatham, Hood and Charles are so different from each other that they are reckoned as separate species, while Swarth (1931) divides the fourth species into seven races, one on Barrington, the second on Indefatigable, Albemarle and Narborough, the third on James and Bindloe, the fourth on Abingdon, the fifth on Tower, the sixth on Wenman and the seventh on Culpepper. The Galapagos mockingbird shows a greater subdivision into island forms than any other Galapagos bird. The differences between the Hood and Indefatigable species are readily apparent, the Hood bird being greyer, less black and white, larger and with a longer and more decurved beak (the Indefatigable mockingbird is shown in Plate VI).

Finally, the most advanced stage of differentiation is shown by Darwin's finches, which are so distinctive that there is considerable doubt as to their nearest mainland relative, while not only are there many island forms, but on each island there are a number of species, up to ten, and even several distinct genera. This is a great increase in complexity beyond the evolutionary stage reached by the Galapagos mockingbird, which has just one form on each island. Nevertheless, the view advocated here is that Darwin's finches passed through the successive stages shown by the other Galapagos land birds, and that they represent simply a much more advanced stage of the same process of differentiation. The steps by which this has come about form the subject-matter of the final chapters of this book.

COCOS LAND BIRDS

One of Darwin's finches, namely, the distinctive *Pinaroloxias inornata*, occurs on Cocos, and this island has also been colonized by two other passerine birds and a cuckoo. The warbler *Den-*

droica petechia is represented by the same race as in the Galapagos, while the other two species are peculiar to Cocos, the cuckoo *Coccyzus ferrugineus* belonging to a mainland genus, and the tyrant flycatcher *Nesotriccus townsendi* being placed in a genus by itself.

ABSENCE OF FOOD COMPETITORS

That Darwin's finches are so highly differentiated suggests that they colonized the Galapagos considerably ahead of the other land birds. Therefore for a period, perhaps a very long period, they were probably without food competitors of other species, while at the same time a variety of foods and habitats was available to them. Even at the present time there are few other passerine birds in the Galapagos, and these are predominantly insectivorous species, some of which may perhaps compete with the warbler-finch *Certhidea*, and possibly with the insectivorous tree-finches, but they compete scarcely, if at all, with the other forms of Darwin's finches. In particular, the later arrivals include no seed-eating, cactus-feeding or wood-boring birds.

The absence of other land birds has had a most important influence on the evolution of Darwin's finches, since it has allowed them to evolve in directions which otherwise would have been closed to them. Finches do not normally evolve into warbler-like or woodpecker-like forms on the continents, because efficient warblers and woodpeckers are already present there. Had a small American woodpecker been established in the Galapagos, it is most unlikely that the woodpecker-finch *Camarhynchus pallidus* could have evolved, since it must, particularly in its early stages, have been far less efficient then a mainland woodpecker, and so would probably have gone under in competition with it. In the same way, had there been a mainland warbler on the islands, it is doubtful whether the warbler-finch *Certhidea* could ever have appeared. There is now an American warbler in the Galapagos, namely *Dendroica petechia*, but this is almost certainly a recent colonist and evidently arrived too late to prevent the evolution of *Certhidea*, which at the present stage of its evolution seems well able to hold its own. Similarly, it is highly doubtful whether *Opuntia* or *Scalesia* could have evolved into tall trees if the Galapagos had been more adequately colonized by mainland trees.

It should be added that, though largely freed from competition with land birds of other types, some of Darwin's finches have come into competition with each other, and this, as considered later, has had highly important evolutionary consequences.

FREEDOM FROM ENEMIES

After their long flight across the ocean, the ancestors of Darwin's finches entered into a land of abundant foods and varied living quarters, unmarred by the presence of competitive neighbours. This avian paradise probably possessed another considerable amenity, namely, complete freedom from enemies. At the present time, Darwin's finches are preyed on by the Galapagos short-eared owl and perhaps also by the Galapagos barn owl, but not, so far as known, by any other animals (see p. 34). Before the time when the owls arrived, Darwin's finches may well have had no enemies at all, and since both owls are closely related to mainland species, they are perhaps fairly recent colonists of the islands.

The absence of predators probably means that Darwin's finches have been limited in numbers primarily by their food supply. When this is the case, adaptations in feeding methods are likely to be of special importance in determining the survival of species, so that the absence of predators may well have accelerated the adaptive radiation of the finches. Similarly, Worthington (1940) has correlated the adaptive radiation of cichlid fish in some of the East African lakes with the absence of predators, in which view he has been supported by Huxley (1942), though Mayr (1942) has brought forward some relevant criticisms of his evidence.

Another effect of the absence of predators is to permit evolution in certain directions which would otherwise be impossible. For example, the Galapagos possess a flightless cormorant *Nannopterum harrissi*. Apart from the ratite birds such as ostriches and rheas, which have evolved speed in running, nearly all the world's flightless birds occur on remote islands free from animals of prey—or rather, formerly occurred there, because many of them have been exterminated in the last few hundred years by men, rats and other predators brought by ships. As already noted, the flight of Darwin's finches is weak and clumsy, and the scarcity of natural enemies has also permitted the remarkable tameness of the finches, though this seems gradually to be disappearing now that they are learning the nature of man.

CHAPTER XIII: THE ORIGIN OF SUBSPECIES

How has it happened in the several [Galapagos] islands situated within sight of each other, having the same geological nature, the same height, climate, etc., that many of the immigrants should have been differently modified, though only in a small degree. This long appeared to me a great difficulty: but it arises in chief part from the deeply-seated error of considering the physical conditions of a country as the most important for its inhabitants; whereas it cannot be disputed that the nature of the other inhabitants with which each has to compete, is at least as important, and generally a far more important element of success....When in former times an immigrant settled on any one or more of the islands, or when it subsequently spread from one island to another, it would undoubtedly be exposed to different conditions of life in the different islands, for it would have to compete with different sets of organisms....If then it varied, natural selection would probably favour different varieties in the different islands.

CHARLES DARWIN: *The Origin of Species*, Ch. XII

DEGREE OF DIFFERENCE SHOWN BY ISLAND FORMS

THE recognition of geographical variation in animals, like the composition of the Star-spangled Banner, came as a by-product of the otherwise unfortunate war of 1812. When replenishing his food supplies with Galapagos tortoises, Captain Porter of the U.S. frigate *Essex* noticed that the specimens from different islands were different in appearance. The point was again noticed by Mr Lawson, the vice-governor of the colony on Charles in 1835, whose name is remembered because he happened to mention the matter to the visiting naturalist of the *Beagle*. The collections then made by Darwin established the existence of distinctive island forms not only in the Galapagos tortoise, but also in the mockingbird *Nesomimus*, in some of the plants such as *Scalesia*, and in some of the finches, though observations on the last group were obscured by his unfortunate mixing of specimens before he became aware of the peculiar state of affairs. Darwin's realization that a species may be represented by different forms in different regions was one of the most important results of the voyage of the *Beagle*, since it led directly to his questioning the immutability of species.

The extent to which Darwin's finches are divided into island forms differs considerably in different species. Thus in the vegetarian tree-finch *Camarhynchus crassirostris*, the birds from different islands show few, if any, differences in either plumage, beak or size. Likewise, the three ground-finches *Geospiza magnirostris*, *G. fortis* and *G. fuliginosa* show no differences in plumage on different islands; but they differ fairly markedly in the average size of beak and wing. An opposite tendency is found in the warbler-finch *Certhidea*, in which the island forms differ markedly in colour of plumage, but less obviously in size of beak and wing. Greater differences occur in the sharp-beaked ground-finch *Geospiza difficilis*, in which there are three clearly defined subspecies differing from each other in plumage, beak, wing-length and ecological niche, while each of these races is in turn slightly different on the two islands on which it occurs. In the cactus ground-finch *G. scandens*, the James and Bindloe forms show differences as marked as those which separate some of the ground-finch species (compare the heads in Fig. 11 (iv), p. 68 and Fig. 12 (i), p. 69). Finally, the large cactus ground-finch *G. conirostris* differs so markedly from *G. scandens* that, though the two forms replace each other geographically, it is doubtful whether they represent pronounced geographical forms of the same original species, or whether they have originated separately; in either case, the differences between them are so great that the birds are classified as separate species.

To sum up, the island forms of Darwin's finches show every stage of differentiation, from differences that are barely perceptible to others that are as marked as those which separate species. These differences are primarily in the size and shape of the beak, the size of the body, and the general shade and the amount of streaking of the female plumage. It is in just these same characters that the individual specimens of one form show differences from each other, and the island populations have presumably diverged from each other by an accumulation and restriction of such individual differences. It is now generally agreed that in animals the differences between geographical races of the same species are hereditary, though in Darwin's finches, as in most other birds, experimental proof of this statement is as yet lacking.

ADAPTIVE AND NON-ADAPTIVE DIFFERENCES

Darwin accounted for the existence of distinctive island forms in the manner quoted at the head of this chapter. As he pointed out, physical and climatic conditions are very similar on the various Galapagos islands, so that there is no reason to think that the differences between island forms are correlated with differences in the physical conditions to which they are subjected. On the other hand, Darwin's postulate that differences in the nature of the competing species are important is abundantly borne out by the data in the latter part of Chapter VI. It was there shown that the small outlying Galapagos islands have a different complement of ground-finch species from the central islands; as a result some species occupy different ecological niches on different islands, in which case the island forms show corresponding differences in the shape of the beak.

There are many other variations between island forms which cannot be related either to differences in the nature of the competing species or to any other differences in the conditions on different islands. This applies to some of the variations in the colour of the female plumage discussed in Chapter IV, to differences in the proportion of fully plumaged males discussed in Chapter V, and to many of the differences in the size of beak and wing discussed in Chapter VII. Both the similarity of the island environments, and the absence of regular trends of variation in the finches, strongly suggest that many of the differences involved are without adaptive significance. As stated earlier, such a negative view is extremely difficult to establish with certainty. There may well be some cases in which an adaptive correlation exists, but has so far been overlooked. But it is extremely unlikely that such a correlation is present but has been overlooked in all of the many cases involved.

A similar problem is presented by the Galapagos reptiles. Many of the islands possess peculiar forms of the tortoise *Testudo*, the snake *Dromicus*, the lizard *Tropidurus* and the gecko *Phyllodactylus*. The ecological niches occupied by these animals seem closely similar on different islands, and it is extremely difficult to believe that the differences in appearance between the island forms are in all cases adaptive. The same applies to many of the differences between the island forms of the Galapagos mocking-bird *Nesomimus*, though, as already noted, the long beak of the Hood form is perhaps adaptive and correlated with its shore-feeding habits.

Unfortunately, the only experimental test of the above views is impracticable, namely, to interchange every single individual of each of two island forms, leaving the two islands and their faunae otherwise unchanged, and then to observe the evolution of each form in the environment of the other over a period of a few hundred, or a few thousand, years. On the view adopted here, there would in many cases be no particular tendency for the transferred island form to evolve the distinctive characters of the previous occupant. But in the absence of such an experiment, this view remains unverified.

IMPORTANCE OF ISOLATION

In continental birds, colour of plumage and average size of beak or wing often show a very gradual change (a cline) over an extensive region, followed by an abrupt change over a narrow region. The regions of rapid change form convenient boundaries at which to delimit geographical races or subspecies, and the narrowness of such zones has been attributed to the fact that hybrids between populations are usually at a disadvantage compared with either parent form (Huxley, 1942). The zones of rapid change tend to occur where intermixing of populations is restricted by geographical obstacles, but they sometimes occur in the absence of the latter, while at the other extreme some continental races are isolated by geographical barriers as completely as are island races. Many of the differences between continental races are undoubtedly adaptive, as mentioned for plumage in Chapter IV and for size in Chapter VII, but in other cases the differences seem as pointless as those which so often separate island races.

Some might claim that all the differences between geographical races of birds are really adaptive; that if any seem pointless, this is due merely to ignorance, and further study will reveal their adaptive significance. But in this connection, a comparison between continental and island races of birds is highly illuminating. Over large land areas the environment shows gradual but marked changes, so that there is every reason to expect continental races to differ from each other adaptively, and in fact they often do so. On the other hand, neighbouring islands often provide very similar environments, so that one might expect any adaptive differences between island races to be much smaller than those separating continental races. Despite this, island races differ from each other much more strikingly than do continental races. This principle is illustrated by almost every bird species whose range includes both continental areas and islands, and the more isolated the island, the greater the degree of differentiation found among its land birds. The primary cause of geographical variation in birds would seem to be not adaptation, but isolation. This point is demonstrated for Darwin's finches in Table XVII (overleaf).

Table XVII and Fig. 22 show that the most isolated islands, namely, Cocos, Culpepper, Wenman, Tower and Hood, have a much higher proportion of peculiar forms than have the central Galapagos islands, while on moderately isolated islands, such as Abingdon, Bindloe, Chatham and Charles, there is an intermediate condition, with proportionately fewer endemic forms than on the remote islands, but proportionately more than on the central islands. Hence in Darwin's finches there is a marked correlation between the degree of isolation and the tendency to produce peculiar forms. In some cases the peculiarities of the forms on the outlying islands are adaptive, being correlated with differences in ecological niche, while in other cases the differences seem unrelated to any possible environmental differences.

The taxonomic review by Swarth (1931) shows that the same principle holds for other Galapagos land birds. For instance, the ground-dove *Nesopelia galapagoensis*, found throughout the archipelago, is represented by a distinctive race on remote Culpepper and Wenman. The mockingbird *Nesomimus* is most distinctive on Hood, Charles and Chatham. The latter islands are more isolated from the bird point of view than a map might suggest, since they lie to the south and south-east, in which direction bird dispersal is comparatively difficult as the trade wind blows from this quarter. The vermilion flycatcher *Pyrocephalus* is also most distinctive on Chatham.

TABLE XVII. ISOLATION AND ENDEMISM IN DARWIN'S FINCHES

Island	Degree of isolation	No. of resident species	Endemic subspecies not found on other islands	
			No.	%
Cocos	Very extreme	1	1	100
Culpepper and Wenman	Extreme	4	3	75
Hood	Marked	3	2	67
Tower	Marked	4	2	50
Chatham	Moderate	7	2½	36
Abingdon and Bindloe	Moderate	9	3	33
Charles	Moderate	8	2	25
Albemarle and Narborough	Small	10	2	20
Barrington	Small	7	1	14
James	Very slight	10	½	5
Jervis	Very slight	9	—	0
Indefatigable	Very slight	10	—	0
Duncan	Very slight	9	—	0

Notes. (i) The number of endemic forms is somewhat arbitrary and is based on Table IV, p. 20. If the classification by Swarth (1931) had been used, the same general correlation would have been apparent, but the percentage of endemic forms would have been higher.

(ii) The forms of *Geospiza scandens* were merged for purely practical reasons of nomenclature (see p. 20), and for assessing insular differentiation the Abingdon and Bindloe birds may be reckoned as distinct, and those from James and Chatham as half-differentiated.

(iii) If Darwin's specimens of *G. magnirostris* and *G. nebulosa* came from Charles (see pp. 22–3), then the figures for Charles should be 10 resident species, 4 endemic forms, or 40 per cent.

(iv) The degree of isolation is assessed from the map. In three cases it was considered best to group a pair of islands together, as they are near each other but distant from all other islands.

Similarly, Murphy (1938*a*) has correlated the degree of isolation and the degree of differentiation in the island forms of various Polynesian land birds, and has shown the important influence of the prevailing wind. Again, Perkins (1913) found that in Hawaii the most isolated islands had the most highly modified birds, and also the greatest number of peculiar insects. The same principle holds in other animals, particularly clear examples being given by Kramer and Mertens for the lizard *Lacerta sicula* on the Adriatic islands, by Kinsey for the gall wasp *Cynips* in North America, and by Reinig for the bumblebee

Bombus in Europe. These latter cases are summarized by Huxley (1942) and Mayr (1942), who also give numerous other examples of geographical variation in animals, the differences being in some cases adaptive, and in others apparently not.

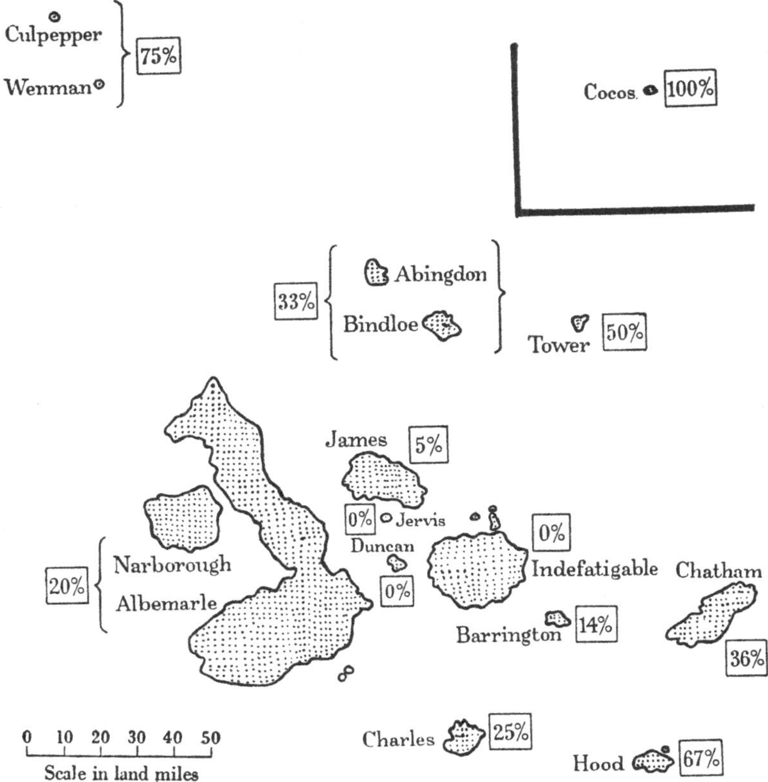

Fig. 22. Percentage of endemic forms of Darwin's finches on each island, showing effect of isolation.

It is at first sight curious that the bird populations of neighbouring islands could be sufficiently isolated from each other to permit the evolution of distinctive forms, since birds are one of the few groups of land animals capable of active dispersal from one island to another. Thus one of Darwin's finches could often reach an island inhabited by a different form by means of an hour's continuous flight, and these birds seem sufficiently strong on the wing to fly such a distance with ease. Mayr (1942) has produced an even more striking case in the white-eye *Zosterops rendovae* of the Solomons, in which one island form could fly into the region inhabited by another form in under five minutes. As Mayr points out, though birds are capable of flying long distances, they tend to use their wings to stay in, or fly back to, their homes, for which reason bird populations are often more isolated from each other than might otherwise have been expected.

EVOLUTION OF NON-ADAPTIVE DIFFERENCES

The evidence considered above shows the fundamental importance of isolation in the differentiation of bird populations, and suggests that many of the differences between the forms are not correlated with differences in their environments. Writers on genetics have suggested several possible mechanisms for the origin of non-adaptive differences between populations. These views have been discussed in detail by Dobzhansky (1937), Huxley (1942) and others, so that only a brief summary is given here.

First, Muller (1940) considers that if two populations are isolated, then through chance alone some of the mutations occurring in one population will be different from those occurring in the other. Such differences lead in turn to further balancing mutations, so that with time the two populations become increasingly divergent. Except as noted in the next paragraph, a mutation must normally be advantageous in order to spread through a population, hence in one sense the differences which arise between the two populations are adaptive; but the point is that such new characters need not be related to possible differences between the environments of the two forms, and they are evolved even if the two environments are identical.

Secondly, Sewall Wright (1940) has shown that if an island population is sufficiently small, of the order of several hundred individuals, then through accident alone slightly favourable mutations may become eliminated and slightly unfavourable mutations established. In this way two small isolated populations of a species may acquire genuinely non-adaptive differences.

Thirdly, a mutation tends to have more than one effect. If a mutation occurred which had a favourable effect, it would tend to spread through a population, though it might carry with it other effects which were neutral in their influence.

Fourthly, one island population might differ from another because the island was originally colonized only by a very few individuals, and these happened not to be typical of the population from which they came. But, at least in birds, the importance of this factor has perhaps been exaggerated. As pointed out by Mayr (1942), bird species tend to have periods of expansion and rapid spread and other periods when they remain more or less static. Most colonization of new regions probably occurs in the periods of expansion, and at such times it seems likely that, if one individual can reach a new locality, quite a number of others will follow. This applies particularly in such regions as the Galapagos, where the distances between islands are comparatively short, in most cases less than 50 miles.

The way in which a bird species becomes established on new ground is being demonstrated at the present time by the colonization of England by the black redstart *Phoenicurus ochrurus*. The data of Witherby and Fitter (1942) show that this species has appeared independently in a number of widely separated localities, while its rate of increase seems greater than can be accounted for solely through the breeding of the existing English stock. Evidently there has been multiple colonization from the European continent, and further individuals are still continuing to arrive. It is therefore probable that the eventual English breeding population will be a typical sample of the stock from western Europe.

Exact counts are not available for the size of the populations of any of Darwin's finches, but rough estimates can be made from the known size of the islands, together with the density of the birds as assessed from field observations. A calculation of this nature suggests that on Daphne, which is only half a mile across, there can exist at one time only a few hundred individuals of the peculiar local form of *Geospiza fortis*, and the Crossman form of *G. fuliginosa* is probably represented by a similar total. Likewise each of the distinctive races inhabiting the islands of Culpepper and Wenman probably consists of only a few hundred, or at most a few thousand, individuals at any one time, and the same holds for the Tower form of *G. conirostris*. Mayr (1942) cites several other island birds with populations of under a thousand individuals. In such cases the existence of marked and non-adaptive differences between island forms might be explained through Sewall Wright's views on the accidental elimination and fixation of hereditary factors in small populations.

But most other island forms of Darwin's finches consist of larger populations than those cited above. On Tower the observed density of *G. magnirostris*, *G. difficilis* and *Certhidea* suggests that there are several thousand individuals of each present there. The other Galapagos islands are considerably larger, and here most forms probably run to tens of thousands, and others to hundreds of thousands, of individuals. Few estimates are available for the size of the populations of other birds, but the great majority of geographical races almost certainly include at least ten thousand individuals alive at any one time, and many include a much greater number. For instance, some of the endemic British song-birds probably include over a million individuals, and some of the continental races even more.

Sewall Wright's views on the accidental elimination and recombination of hereditary characters are not applicable to populations consisting of more than a few hundred individuals. Hence they cannot be invoked to explain the differences between geographical races of birds except in a small minority of cases. It might, of course, be contended that Darwin's finches decrease periodically to numbers much smaller than those given above, but there is no evidence for this, and there are many other bird races whose populations almost certainly never fall as low as a thousand individuals. Perhaps the views of Muller, summarized earlier, are sufficient to account for the differences in appearance between these larger populations, but this subject requires much further study. To conclude, the evidence from Darwin's finches and other birds shows the great importance of geographical isolation in producing hereditary differences between populations, but there is still considerable doubt as to the way in which they are actually brought about.

CHAPTER XIV: THE ORIGIN OF SPECIES

No clear line has as yet been drawn between species and sub-species...or, again, between sub-species and well-marked varieties, or between lesser varieties and individual differences. These differences blend into each other in an insensible series; and a series impresses the mind with an actual passage.

CHARLES DARWIN: *The Origin of Species*, Ch. II

INCIPIENT SPECIES

THE various specimens of Darwin's finches from any one island do not form a continuously graded series from large to small, thick-billed to thin-billed, or dark to pale. Instead they fall into distinct segregated groups, each group having a characteristic appearance, while the individuals of any one group do not normally interbreed with the members of any other. A similar state of affairs is found in the birds breeding in Britain, and for that matter in every other region of the world. The segregated groups are, of course, the units termed species, and their manner of origin has aroused discussion and controversy ever since the theory of evolution came to be accepted.

The apparent fixity of species is most striking, and provides the basis for systematic zoology. But with the full acceptance of the doctrine of evolution there has arisen a tendency among general biologists, though not among taxonomists, to underestimate the definite nature of species, and a corresponding tendency to exaggerate the frequency of intermediate forms. Charles Darwin and many after him are partly wrong when they assert that the determination of species is purely arbitrary. Provided the ornithologist keeps within a limited district, he is usually in no doubt as to which birds should be regarded as separate species, and the same holds in many other groups of animals. Difficulty over intermediate forms arises mainly when the naturalist compares related forms of birds or other animals from different districts, but then the difficulty immediately becomes considerable, a fact which provides the essential clue to the way in which new species originate.

Big evolutionary changes are normally achieved in a series of small steps, so that it is to be expected that the gaps between species would come into existence gradually, in which case some of the intermediate stages ought to be visible. The closely related species of Darwin's finches differ from each other in beak, in size of body, in the shade and amount of streaking of the female plumage, and in the amount of black in the male plumage. It is in just these characters that island forms of the same species differ from each other and, as discussed in the last chapter, such island forms show every stage of divergence from differences that are barely perceptible to differences as marked as those which separate some of the species. Moreover, this is the only kind of incipient differentiation found among Darwin's finches. These facts strongly suggest that island forms are species in the making, and that new species have arisen when well-differentiated island forms have later met in the same region and kept distinct.

CAMARHYNCHUS ON CHARLES

An instance of such a manner of origin is provided by the two species of large insectivorous tree-finch, *Camarhynchus psittacula* and *C. pauper*, which occur together on Charles. These are undoubtedly two separate species, differing in size of beak, wing-length and shade of plumage. The differences between them are small, but constant and reliable, and each collected specimen can safely be allocated to one or the other type.

If *C. psittacula* (*sens. strict.*) did not occur on Charles, all the large insectivorous tree-finches could be included in one species *C. psittacula*, divided into four well-marked geographical races as follows: *psittacula* (*sens. strict.*) on the central islands of James, Indefatigable and Barrington, *habeli* to the north on Abingdon and Bindloe, *affinis* to the west on Albemarle and Narborough, and *pauper* to the south on Charles, as shown in Fig. 23. Of these four forms the Charles form *pauper* appears to be the most primitive, being much streaked and possessing the smallest and most finch-like beak. The Albemarle form *affinis* shows close resemblance to *pauper* both in plumage and beak, so links up with it. The central island form *psittacula* (*sens. strict.*) is less streaked, larger and with a more parrot-shaped beak, but links up with the Albemarle form through a population of intermediate type on the intervening island of Duncan. The northern form *habeli* shows most resemblance to *psittacula* (*sens. strict.*).

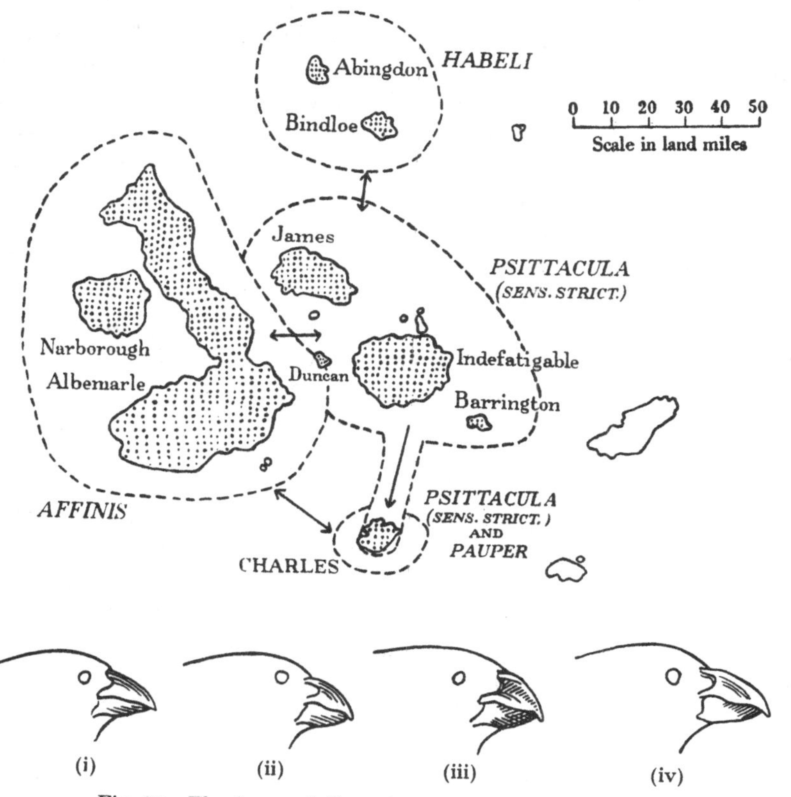

Fig. 23. The forms of *Camarhynchus psittacula* (*sens. lat.*).

Showing the existence of two forms on Charles, the earlier form (*C. pauper*) being related to the Albemarle form (*affinis*), and the later form (*psittacula, sens. strict.*) coming from the central islands. The two have met, but do not interbreed.

(i) *pauper* (Charles) (ii) *affinis* (Albemarle) (iii) *psittacula, sens. strict.*, (James) (iv) *habeli* (Bindloe)

Heads ⅞ natural size (*after* Swarth).

This simple situation is complicated by the fact that Charles is inhabited not only by the form *pauper* but also by the form *psittacula* (*sens. strict.*), the Charles individuals of the latter being indistinguishable from those found on the central islands. So far as known these two forms do not interbreed on Charles, and there are no specimens intermediate between them in appearance. The facts suggest that Charles has been colonized by the large insectivorous tree-finch on two separate occasions. Originally, the island was inhabited by the form *pauper*, or by a form which later turned into *pauper*, while more recently it has been invaded from the north by the form *psittacula* (*sens. strict.*). Formerly, *pauper* and *psittacula* (*sens. strict.*) were geographical races of the same species, but by the time that they met on Charles they had become so different that they did not interbreed, and so they have become separate species.

Similarly, if the Albemarle race *affinis* were now to colonize Abingdon, where the form *habeli* lives, the two are so distinctive that they might keep separate, in which case they also would have to be classified as separate species. But such an invasion has not yet occurred, so that it is more convenient to consider *affinis* and *habeli* as races of the same species.

SPECIES-FORMATION IN OTHER BIRDS

Darwin's finches provide no further cases in which the origin of a new species can be traced in this way. But during the last ten years many similar examples have come to light in other birds. These have been fully reviewed by Mayr (1942), so that detailed comment is scarcely necessary here. A close parallel with the case of *Camarhynchus* on Charles is provided by the two species of thornbill *Acanthiza ewingii* and *A. pusilla* in Tasmania. On the Australian mainland the species *A. pusilla* is divided into a number of well-defined geographical races, each occupying a particular region. But on Tasmania two related forms, both obviously derived from *pusilla* stock, occur together in the same region without interbreeding. They must therefore be reckoned as separate species, though they differ no more in appearance than do some of the mainland races of *pusilla*. Presumably the original Tasmanian form *ewingii* had become so distinctive that, when a second invasion of *pusilla* took place from Australia, the two forms kept separate.

Mayr gives a number of other examples in which a remote island is inhabited by two closely related species, both evidently derived from the same mainland species, and which have presumably originated as a result of two separate colonizations of the island by the mainland form. More comparable with the situation in Darwin's finches are several cases in which an island appears to have been colonized on two separate occasions from adjoining islands. This, for instance, probably accounts for the existence of two related species of the finch *Nesospiza* on the same islands in the Tristan da Cunha archipelago, for the two species of the flycatcher *Mayrornis* on Ongea Levu in the Fijis, for the two species of the fruit pigeon *Ptilinopus* in the Marquesas, and other cases.

There are also a number of comparable examples among continental birds. For instance, the herring gull *Larus argentatus* and the lesser black-backed gull *Larus fuscus* both breed in Britain, and they do not usually interbreed. But if they had not thus met in western Europe, they might have been classified as geographical races of the same species, as they are linked by a series of geographical forms extending across Europe, Asia and North America. By the time that this species had spread right round the world, the two end-forms had evidently become sufficiently different not to interbreed where they met. A similar situation is presented by the two species of great tit *Parus major* and *P. minor*. These species occur together in the Amur valley, where they keep distinct, but each is linked by a chain of geographical races with the same great-tit stock. Mayr gives other examples.

PREVENTION OF INTERBREEDING

In birds generally, as in Darwin's finches, geographical forms show every stage of divergence, from differences which are barely perceptible to differences as marked as those which separate full species. From the evidence which has now accumulated, it is clear that the commonest method of species-formation in birds is through the meeting in the same region of two geographical forms which have become so different that they keep separate. The fundamental problem in the origin of species is not the origin of differences in appearance, since these arise at the level of the geographical race, but the origin of genetic segregation. The test of species-formation is whether, when two forms meet, they interbreed and merge, or whether they keep distinct.

As discussed in the last chapter, when two populations of the same form are isolated from each other, differences gradually arise between them. Muller (1940) considers further that such differences inevitably lead to some degree of sterility between the individuals of the two populations. If this view is correct, there should sometimes be partial sterility between geographical races of the same species, and this has now been established in the case of a number of insects, as summarized by Mayr (1942). Similar evidence is not yet available in birds, as they are rather unsuitable for quantitative breeding experiments.

If the members of two well-differentiated races meet later in the same region, and if they are partially intersterile, or if their hybrid offspring are at a disadvantage, then those individuals which breed with members of their own kind tend to leave more offspring than those which interbreed with individuals of the other race. Hence even if genetical segregation between two races is not complete when they first meet, natural selection will tend to deepen the gap between them. Indeed, it has even been claimed that, owing to the disadvantages possessed by hybrids, selection will initiate intersterility between forms which meet in this way. However, the latter view is not certain.

The above considerations show that any factors which prevent the interbreeding of forms have survival value, hence the frequency with which specific recognition marks have been evolved in birds, as discussed in Chapter v. Darwin's finches are unusual in that the beak is used as a recognition mark, but as this is the most prominent racial and specific difference, it is not surprising that the birds should have evolved behaviour responses relating to it.

Barriers to interbreeding might also be provided by differences in breeding season or habitat. But the various species of Darwin's finches breed at the same season and most of them are not separated in habitat, so that these factors have little or no importance. The latter conclusion applies to birds generally Most related species found in the same region breed at the same season, and though they commonly occupy different habitats, the degree of isolation thus provided is usually quite inadequate to ensure genetic isolation. In other birds, as in Darwin's finches, the primary factors which prevent attempts at interbreeding are psychological ones correlated with breeding behaviour.

OTHER METHODS OF SPECIES-FORMATION

Darwin's statement quoted at the head of this chapter suggests that in animals every gradation exists between mere varieties and full species. Later knowledge has shown that, at least in birds, this statement is somewhat misleading. There is only one kind of variety, if such it should be called, which grades insensibly with the full species, namely the geographical race. In birds there is no other type of variety which can reasonably be termed a subspecies, indeed, the terms subspecies and geographical race have become synonymous. This strongly suggests that in birds the only regular method of species-formation is via races differentiated in geographical isolation. However, it has sometimes been claimed that, to produce the variety of species found in Darwin's finches, some quite peculiar method of evolution must have been involved. Even Rensch (1933), who was the first to advocate species-formation from geographical races as a widespread principle, was greatly puzzled by Darwin's finches, and considered that for them some different process must have operated.

The frequency with which closely related bird species occupy different habitats suggests that an alternative method of species-formation is by ecological, instead of geographical, isolation. Since a bird tends to breed in the same type of habitat as that in which it was raised, it might have been expected that, where a species breeds in a variety of habitats, it would tend to become subdivided into populations each with a rather different habitat preference, and that with time this might result in the formation of new species. This is a plausible view and has been put forward by a number of writers, formerly including myself (1933). But there are two insuperable objections. First, the degree of isolation provided by differences in habitat is not usually at all complete, and the bird species which occupy separate habitats usually have numerous border zones where they come in contact with other species. To produce well-differentiated forms, complete isolation seems essential. Secondly, no cases are known in birds of incipient species in process of differentiation in adjoining habitats. All subspecies are isolated from each other geographically and, though they occasionally differ in habitat as well, geographical isolation is the essential factor. These points are treated in further detail by Mayr (1942). Finally, the frequent existence of habitat differences between closely related bird species has a quite different explanation, consideration of which is postponed to the next chapter.

Another alternative has been suggested by Lowe (1930, 1936), who supposes that Darwin's finches represent the varied products of interbreeding between a small number of original forms, as has happened in certain 'species-swarms' in plants. The reasons for rejecting this view have been discussed in Chapter x. There is no evidence that hybridization has been of importance in species-formation in any group of birds.

Streseman (1936) is the only previous writer to suggest that species-formation has followed the same course in Darwin's finches as in other birds, i.e. that forms differentiated in geographical isolation have later met and kept distinct. With this conclusion, I fully agree. There is only one apparent case, in the large insectivorous tree-finches *Camarhynchus psittacula* and *C. pauper* on Charles. But such cases will rarely be apparent, since once a form has become firmly established in the range of another it will tend to spread rapidly right through that range, so that its place and means of origin quickly become obscured. The only type of incipient differentiation found in Darwin's finches is that shown by geographical races, and there is nothing to suggest that geographical isolation is not the essential preliminary to species-formation in this group. The existence of an unusually large number of similar species may be attributed first to the great length of time for which the finches have been in the Galapagos, secondly to the paucity of other land birds, and thirdly to the unusually favourable conditions provided by a group of oceanic islands, both for differentiation in temporary geographical isolation, and also for the subsequent meeting of forms after differentiation.

The primary importance of the geographical factor is strongly corroborated by the situation on Cocos Island. Here there occurs one, and only one, species of Darwin's finch, *Pinaroloxias inornata*. That it has been on Cocos a long time is suggested by the extent to which it differs from all the other species of Darwin's finches. Yet despite the length of time for which it has been there, despite the variety of foods and habitats which Cocos provides, and despite the almost complete absence of both food competitors and enemies, there is still only one species of Darwin's finch on Cocos. But Cocos is a single island, not an archipelago, and so provides no opportunity for the differentiation of forms in geographical isolation.

SPECIES-FORMATION IN OTHER ORGANISMS

That geographical isolation can lead to the origin of new species in groups other than birds is shown by examples given by Mayr (1942) and Huxley (1942) for mammals, reptiles, amphibia, molluscs and several groups of insects. Populations of organisms such as insects or land molluscs are sometimes isolated from each other in a much smaller space than are bird populations. It might therefore be better to replace the term geographical isolation by topographical isolation when considering animals generally.

The extent to which ecological isolation can lead to the formation of new species in other animals is still in doubt (Mayr, 1942; Huxley, 1942, modified in 1943). If an insect occurred on several types of food plant, or if a parasite had several host species, it seems possible that populations might become isolated, each restricted to a particular food plant or host species, thus allowing subspecific and eventually specific differentiation to take place. But this is not proven and, except in these rather special circumstances, it seems doubtful whether ecological differences without topographical isolation can provide a sufficient degree of segregation between populations. There are many groups of animals and plants in which the closely related species differ in their ecology, but it does not necessarily follow that ecological isolation preceded the formation of the species in question, as will become clear in the next chapter.

In certain types of plants big changes in the chromosomes can produce sterility between individuals, and it is possible that such changes can lead to the effective isolation of groups within the species, and so to the formation of new species. Such purely genetic isolation may also have been important in a few animal groups but, as Huxley (1942) points out, in most animals purely genetic isolation becomes important only secondarily, after initial differentiation in topographical isolation. A different type of genetic isolation is found in the few plants and the much smaller number of animals which reproduce only by asexual means, and such forms are often highly differentiated.

To conclude, in all organisms the isolation of populations is an essential preliminary to the origin of new species. In birds, geographical isolation is of primary importance, and isolation by genetic factors arises secondarily; ecological isolation is not known to initiate species-formation. Possibly these conclusions apply to many other groups of animals and also to plants, but definite conclusions cannot be reached until the systematics of other organisms is known as well as that of birds, and, in particular, the systematics of the subspecific units which are species in the making.

Chapter XV: THE PERSISTENCE OF SPECIES

If a variety were to flourish...it might come to supplant and exterminate the parent species; or both might co-exist, and both rank as independent species.

Charles Darwin: *The Origin of Species*, Ch. II

WHEN SPECIES MEET

Closely related species of animals often differ from each other only in small and apparently trivial ways. After a careful survey of the evidence at that time available, Robson and Richards (1936) concluded that the differences between such species are not usually adaptive, and that adaptive differences tend to appear only at the level of divergence represented by the genus or subgenus. This view has been widely accepted, and, if true, constitutes one of the most puzzling features of the species problem. But further consideration has led me to realize that the absence of adaptive differences is only apparent, and that in fact closely related species differ from each other in ways which play an extremely important part in determining their survival.

When two forms, originally geographical races of the same species, meet later in the same region and keep distinct, thus forming separate species, there is raised not only an important genetical problem, the prevention of interbreeding, but an even more important ecological problem, since the two forms will tend to compete against each other. The chance is negligible that, after their differentiation in isolation, both forms should be equally efficient in every respect, so that the following possibilities exist.

First, one of the two forms may be so much better adapted than the other that it spreads rapidly right through the range of the other and exterminates it. This seems the most likely possibility, but will rarely be observed as it leaves no trace.

Secondly, one form may be better adapted than the other in the region where they meet but, after it has eliminated the less successful form from part of its range, it may come to a region where the environment is better suited to the other form. In this case the two species will come to occupy separate but adjoining geographical regions. Since environmental factors tend to change gradually, there may be a region where both forms are about equally well adapted, and here their ranges will overlap.

Thirdly, one form may prove better adapted to one section of the original habitat, and the other to the rest. In this case each will tend to spread through the geographical range of the other, each eliminating the other from part of its original habitat, so that they come to occupy separate but adjacent habitats in the same regions.

Fourthly, one form may prove better adapted for obtaining certain foods, the other for taking other foods In this case, if their numbers are limited primarily by food supply, the two species may be able to co-exist in the same habitat, dividing the available foods. The evidence discussed in Chapter VI suggests that in such cases the foods taken by the two species need be only partly and not wholly different, and that a difference in food habits is commonly associated with a marked difference in size, including size of beak.

Particularly in migratory birds, the above possibilities may be modified by the seasonal factor. Two species need not be isolated from each other in the same way at all seasons; for instance, they might occupy different habitats in the same region when breeding, but a similar habitat in separate regions in winter. Further, if mutual competition would seriously affect their numbers at only one time of year, for instance, in the breeding season, they might be able to mix freely at other seasons without effectively competing. Another but less likely possibility is that two species might inhabit the same place, but breed at different times of year. It is also possible for two species to be isolated in different ways in different parts of their range, in one part separated in habitat and elsewhere geographically.

Dr G. C. Varley has pointed out to me that the above possibilities apply primarily to animals which are limited by food supply. In the case of species whose numbers are controlled by parasites or predators, the population density may be greatly below the limit set by food, so that two species could live in the same habitat and eat the same foods without effectively competing. This type of situation seems much more likely to be important in animals such as insects than in birds.

ECOLOGICAL ISOLATION IN DARWIN'S FINCHES

Darwin's finches provide considerable support for the correctness of the above views, since all the closely related species appear to be isolated from each other in one way or another. This is shown in the following summary, which is based on the data given previously in Chapters II, III and VI. The sharp-beaked ground-finch *Geospiza difficilis* is of particular interest, as on the central islands it differs from the small ground-finch *G. fuliginosa* in habitat, while on the northern islands it is separated from it geographically.

Geographical Separation: 2 cases

(i) The cactus ground-finch *G. scandens* occurs on most islands, but not on Hood, Tower or Culpepper. The large cactus ground-finch *G. conirostris* occurs only on these three latter islands.

(ii) The small ground-finch *G. fuliginosa* inhabits the arid zone of most islands, but is absent from Culpepper, Wenman and Tower. Only on these three latter islands does the sharp-beaked ground-finch *G. difficilis* occupy the arid zone.

Separation by Habitat: 2 cases

(i) On the central islands *G. difficilis* breeds only in the humid forest, and *G. fuliginosa* only in the arid and transitional zones. (See also preceding section.)

(ii) On Albemarle and Narborough, the mangrove-finch *Camarhynchus heliobates* breeds only in the coastal mangrove belt and the woodpecker-finch *C. pallidus* inland. (These two species are in the same subgenus.)

SEPARATION BY FEEDING HABITS: 2 cases

(i) The cactus ground-finch *Geospiza scandens* breeds in the same habitat with the other ground-finches, *G. magnirostris*, *G. fortis* and *G. fuliginosa*, but its chief food is *Opuntia*, which they do not normally eat.

(ii) The vegetarian tree-finch *Camarhynchus crassirostris* feeds chiefly on leaves, fruits and buds, whereas the other species of *Camarhynchus* are mainly insectivorous.

SEPARATION BY SIZE OF BEAK, AND PRESUMABLY BY FOOD: 4 cases

(i, ii) The large ground-finch *Geospiza magnirostris*, the medium *G. fortis* and the small *G. fuliginosa* occupy the same habitat but differ markedly in size of beak; their foods are partly different.

(iii, iv) A similar situation is presented by the insectivorous tree-finches, the large *Camarhynchus psittacula* and the small *C. parvulus*, with a third species of medium size, *C. pauper*, on Charles. The foods of these species have not been analysed.

ECOLOGICAL ISOLATION IN OTHER BIRDS

A parallel survey of British passerine birds (Lack, 1944*a*) shows that in these birds, as in Darwin's finches, closely related species are normally isolated from each other ecologically. In some cases they breed in separate geographical regions, like the carrion and hooded crows *Corvus corone* and *C. cornix*. In other cases they occupy separate breeding habitats, like the meadow, tree and rock pipits, *Anthus pratensis*, *A. trivialis* and *A. spinoletta petrosus*. In yet other cases two species breed together in the same habitat, but under these circumstances they normally take different foods. Thus the spotted and pied flycatchers *Muscicapa striata* and *M. hypoleuca* have rather different methods of catching their food; the chiffchaff and willow warbler *Phylloscopus collybita* and *P. trochilus* tend to feed at different levels in the woods, the former higher than the latter; and the corn and yellow buntings *Emberiza calandra* and *E. citrinella* differ conspicuously in size and in size of beak, and by inference in their foods. In some of these cases the birds may be separated in a different way in winter. For instance, the tree and meadow pipits then occupy separate geographical regions, and the same applies to the chiffchaff and willow warbler.

A quick walk through the English countryside might suggest that there was wide ecological overlap between the various songbirds. In fact, close analysis shows that there are extremely few cases in which two species with similar feeding habits are found in the same habitat. One apparent example is provided by the blackcap and garden warbler *Sylvia atricapilla* and *S. borin*, both of which breed in woods with good secondary growth. Possibly this and the few other cases of apparent ecological overlap are due merely to inadequate observation.

Parallel with the survey of British birds, a review was made of those instances in which two closely related passerine species occur together on the same remote island (Lack, 1944*a*). In these cases also, the two species normally have different ecology. Instances in which they differ markedly in size and in size of beak were mentioned in Chapter VI, and two of these cases, that of the white-eye *Zosterops* on Lord Howe Island and the finch *Nesospiza* on Nightingale Island, Tristan da Cunha, are shown in Fig. 24.

Further examples from remote islands need not be considered here, except for an instructive case involving separation by habitat. In the Canary Islands occur two species of chaffinch, the blue chaffinch *Fringilla teydea* and a local form of the European chaffinch *F. coelebs*. Mayr (1942) follows Stresemann in presuming that *F. teydea* was originally a geographical form of *F. coelebs* which became so distinctive that, when the Canary Islands were invaded by the European chaffinch for the second time, the two kept separate and so formed distinct species. On the European mainland *F. coelebs* is widespread in both broad-leaved and coniferous woodland. But in Gran Canaria and Tenerife the blue chaffinch *F. teydea* breeds only in pine forest, while the local form of *F. coelebs* breeds only in the chestnut and laurel forest below the pine belt and in the tree-heath zone above it. On the island of Palma the blue chaffinch is absent, and here, unlike other islands, the local form of *F. coelebs* breeds not only in chestnut and laurel forest but also in pine forest (Bannerman, unpublished). Presumably, when the two chaffinch species met in the Canaries, the blue chaffinch proved better adapted to the pine forest and the newer form to broad-leaved woodland and tree-heath, with the result that they divided the original chaffinch habitat between them. Only where one of the forms is absent, as on Palma, can the other persist in all types of woodland, like its European ancestor.

Cases of ecological isolation through differences in food, habitat or geographical region are common in birds. Isolation through a difference in breeding season is much rarer. No instances are found in Darwin's finches or in British birds, but a

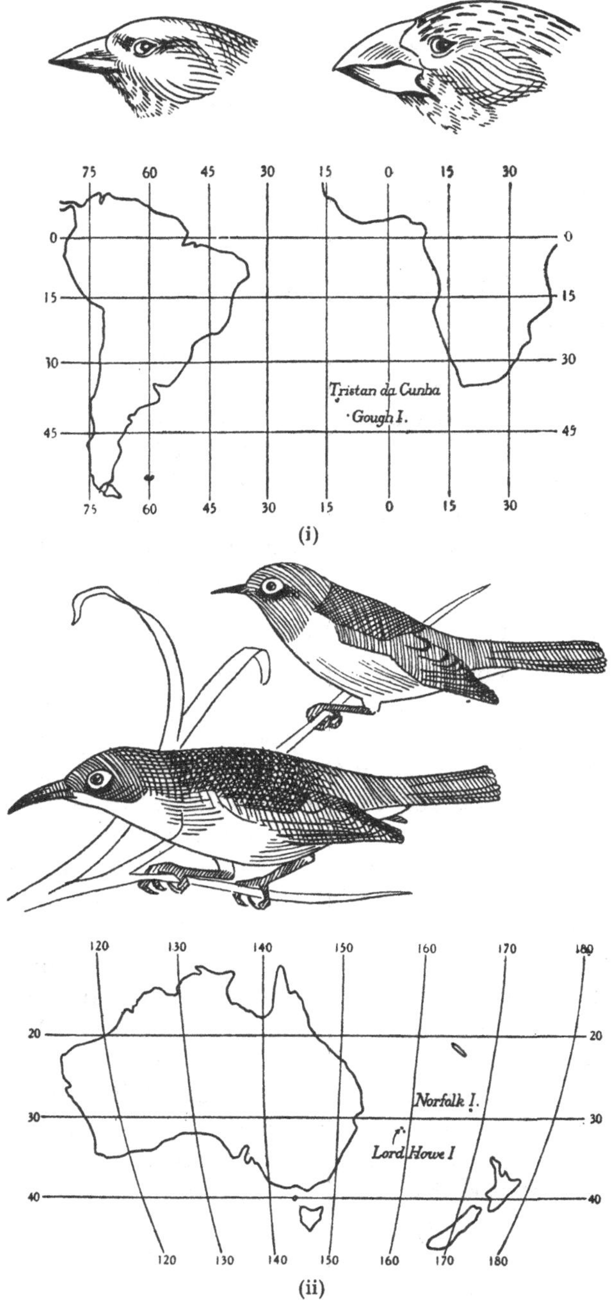

Fig. 24. Two species on the same island differing markedly in size. (i) *Nesospiza* on Tristan da Cunha (ii) *Zosterops* on Lord Howe I (*after* Grönvold)

few are known in sea birds. For example, two species of terns breed on Kerguelen Island, *Sterna virgata* in October and November, while in January and February the same nesting grounds are taken over by *Sterna vittata*. These two species also differ in food, the former feeding on spiders and insects on the inland marshes, the latter being exclusively marine (Murphy, 1938*b*).

These cases, with many others given elsewhere (Lack, 1944*a*), provide strong support for the contention by Gause (1934) that two animal species with the same ecology cannot live in the same region. The reason now becomes clear why closely related passerine birds so often frequent different habitats. This is primarily a result of competition between them. The rarity of habitat differences between species in Darwin's finches is because, in these birds, ecological divergence has mainly taken the form of differences in food habits, in which case the species are not forced to occupy separate habitats, but can exist side by side.

ECOLOGICAL ISOLATION IN OTHER ANIMALS

Birds are not, of course, the only animals in which closely related species either differ in ecology or mutually compete. As mentioned earlier (p. 28), the European bank vole *Clethrionomys glareolus* has largely replaced the related species *C. nageri* where the two came in contact, except at higher altitudes and in the north, where *C. nageri* proves more efficient. Barrett-Hamilton and Hinton (1913–14) state that these two species were originally geographical forms of the same species. Similarly, in Scotland the mountain hare *Lepus timidus* occupies the high ground and the common hare *L. europaeus* the low ground. Fossil evidence shows that a form of the mountain hare was at one time widespread on low ground in Britain. However, it became extinct in most regions after the arrival of the common hare. But Ireland was cut off from the rest of Britain before the common hare could get there, and the Irish form of the mountain hare occupies not only the high ground but also the low ground, suggesting that the absence of the mountain hare from the low ground of Scotland and England is due to competition with the common hare. Barrett-Hamilton and Hinton (1913–14) give another parallel instance of competition and replacement in the British field voles *Microtus corneri*, *M. hirtus* and *M. agrestis*.

In mammals, unlike birds, fossil remains are sometimes available to show that extinction has taken place.

Mammals also provide good examples of species which live in the same habitat but differ in feeding habits. For instance, four species of seals frequent the Ross Sea in the Antarctic. According to Wilson (1907), the crabeater seal *Lobodon carcinophagus* feeds exclusively on euphausiid crustaceans, the Weddell seal *Leptonychotes weddelli* mainly on fish, the leopard seal *Stenorhinchus leptonyx* mainly on penguins and seals, also on carrion, fish and cephalopods, and the Ross seal *Ommatophoca rossi* (of which few specimens are available) mainly on cephalopods. Instances are also found where two closely related species live in the same habitat and differ markedly in size, and by inference in food. Cases in Britain are the common and pygmy shrews *Sorex araneus* and *S. minutus*, and the stoat and weasel *Mustela ermineus* and *M. nivalis* (Huxley, 1942). Hence all the chief types of ecological isolation found in birds are also found in closely related species of mammals.

Turning to a very different type of animal, in Europe the flatworms *Planaria montenegrina* and *P. gonocephala* occupy different parts of the same streams, the former species occurring in waters below a temperature of 13–14° C., and the latter in waters above this temperature. The habitat limitation of each species is not a direct effect of temperature, but is determined by competition between them, since in streams where one of them is totally absent, the other has a wider temperature range. Similar competitive relations hold between two other species of flatworms, namely *Planaria alpina* and *Polycelis cornuta*, in this case the rate of flow of the water being the critical factor (Beauchamp and Ullyott, 1932). In these flatworms, as in birds, habitat differences are the result of inter-specific competition, and the habitat limits of each species are determined by differences in adaptive efficiency in regard to particular environmental factors.

The exclusion of one species by a related species has been shown experimentally by Mayne and Young (1938), who found that the blood of a man can normally support only one species of malarial parasite at one time. If two species, for instance, *Plasmodium malariae* and *P. vivax*, are injected together, only one of them normally persists. A similar result was found by

Gause (1934), in the laboratory cultures of micro-organisms which he used to demonstrate his original thesis that two species with similar ecology cannot live together. The same author cites the gradual replacement of the crayfish *Potamobius astacus* by the related species *P. leptodactylus* in many Russian waters during recent years.

There are many groups of insects in which the closely related species are restricted to different food plants. Numerous examples could be cited from British insects, and Perkins (1913) was particularly impressed with this point in his study of the endemic cerambycid beetles and homopterous bugs of the Hawaiian Islands. However, there are also many instances in which closely related insect species occur on the same food plant; as suggested earlier (p. 136), this is understandable in cases where the populations are controlled primarily by parasites or predators.

To follow the problem of ecological isolation further would require a book in itself. Huxley (1942, pp. 265–84) may be referred to for many additional examples drawn from a great diversity of both animals and plants. These were cited to support his view that ecological isolation is a cause of species-formation (see, however, Huxley, 1943), but they are equally explicable through competition between species. There may well be other groups in which the competitive relations between species are more intricate than in birds, and the subject requires intensive study in all types of organisms. Particular attention might be directed to those cases in which two related species appear to live together in the same region without effectively competing.

ADAPTIVE DIFFERENCES BETWEEN SPECIES

To return to birds, it is clear from the foregoing discussion that, though adaptive differences between species may not be obvious, they must exist. There is no other way of accounting for the ecological isolation of each species. The nature of the adaptive differences presumably depends on the mortality factors which determine whether one or another species is to survive in a particular place. If food supply is limiting, then adaptations for feeding may be paramount, as in Darwin's finches. If predators are particularly important, then adaptive differences may concern means of escape, such as speed of flight or protective coloration. If many of the birds are killed by a climatic factor, then a migratory habit, or resistance to cold, or to drought, may assist in determining which species will persist in a particular region.

In a few cases, closely related bird species are known to differ adaptively. Thus when two species share the same habitat but differ in food, they often differ markedly in size of beak, as discussed in Chapter VI. An instance of an adaptive difference between species which occupy separate geographical regions is provided by the two Atlantic species of guillemot or murre. Brünnich's guillemot *Uria lomvia* is more northerly and also larger than the common guillemot *U. aalge*, a correlation which is in agreement with Bergmann's rule (see p. 79). An adaptive difference correlated with a difference in habitat is provided by the British pipits, the tree pipit *Anthus trivialis* having, like other tree-frequenting birds, a hind claw which is short as compared with that of the meadow pipit *A. pratensis*.

That adaptive differences are not known in a much larger number of cases is probably due simply to inadequate study. For example, it is not known what differences between the ground-finches *Geospiza fuliginosa* and *G. difficilis* make for the success of the former in the arid zone and the latter in the humid forest on the central Galapagos islands. But the distribution of these birds is explicable only on the view that such adaptive differences exist. Likewise adaptive differences must exist between the two species of chaffinch in the Canary Islands, though in this case also, the nature of the differences is not obvious.

Where, as in the above instances, two bird species occupy separate habitats, there is no reason to think that competition between them is direct, meaning that one forcibly drives the other out. Even the most territorial birds attack members of other species only sporadically and ineffectively. A partial exception is provided by the severe competition for nesting sites often found between hole-nesting species, but this is a special case, and is not comparable with driving another species from an entire habitat. Probably the habitat differences between two species are normally brought about gradually by natural selection. Individuals of the one species survive better in those places where the other is at a comparative disadvantage, and vice versa, so that gradually each evolves a specific habitat preference.

In plants, as in birds, closely related species often occupy different habitats, and there is often a similar degree of difficulty in detecting adaptive differences between the species. But in plants the existence, though not the nature, of such differences is readily proved, for if one species is transplanted to the natural habitat of the other, it is usually eliminated. A parallel experiment under natural conditions is impracticable with birds, for on release they simply fly back to their natural homes.

While Robson and Richards (1936) seem mistaken in supposing that closely related species do not differ adaptively, they are correct to the extent that adaptive differences have not, in most cases, been described, and that the characters used by systematists to distinguish related species usually seem to be without adaptive significance. To the latter a prominent exception is provided by the specific recognition marks of many birds, also by the marked size differences often found between related species living in the same habitat. But many other specific differences seem very trivial, both in birds and other animals. That all such differences will eventually prove to be adaptive seems unlikely, but considerable further study is required before a conclusion can safely be reached on this point.

PRE-ADAPTATION

On Lord Howe Island, as already mentioned, occurred two species of white-eye, differing in size of beak and to some extent in food, and apparently derived from separate invasions of the island by the same Australian species *Zosterops lateralis* (see Fig. 24; also Mathews, 1928; Stresemann, 1931; Hindwood, 1940; Lack, 1944*a*). Presumably the earlier form had already become so large that, when the second arrived, the two did not compete sufficiently for one to eliminate the other. The size difference may well have been intensified later by natural selection, but some difference must have been present when they first met. In the same way the earlier form of chaffinch in the Canary Islands must have been better adapted to the pine zone, and the later arrival to the broad-leaved zone, before the period at which the later form colonized. Otherwise one would presumably have eliminated the other completely.

There is nothing difficult in the idea of such pre-adaptation. Though in many cases geographical races of the same species have similar ecology and are of similar size, cases in which they have diverged in these respects are not difficult to find. For instance, races which differ prominently in size of beak are not uncommon in insular birds, and Darwin's finches provide several examples, such as the three races of the sharp-beaked ground-finch *Geospiza difficilis* and those of the large cactus ground-finch *G. conirostris*, discussed in Chapter VI. An example from a continental area is the crossbill *Loxia curvirostra* (p. 80). The ground-finch *G. difficilis* also provides an example of a racial difference in habitat, this, together with other cases in both insular and continental birds, being discussed in Chapter III (pp. 26–30). It is presumably only such highly differentiated forms which are capable of surviving together should they meet in the same region.

Ecological differences between races of the same species seem relatively more frequent on remote islands than on continents. Probably this is because on the continents the chief ecological niches are already filled, and each species is kept to a restricted niche by the presence of other species. Such other species are likely to be present over wide regions, so that there is comparatively little opportunity for the continental races of a species to vary in their ecology. But owing to the difficulties of colonization, comparatively few species are present on remote islands. In particular there were perhaps no other land birds in the Galapagos when Darwin's finches first arrived. Under these circumstances, there are unusually great possibilities of ecological divergence between island forms of the same species. As a result, when two such forms meet later in the same region, there is an unusually good chance that they will be sufficiently different ecologically for both to persist. This is a highly important reason, additional to those considered in the last chapter, for the large number of species into which Darwin's finches have diverged.

To conclude, this and the previous chapter are summarized by saying that, when forms differentiated in geographical isolation meet later in the same region, two conditions are necessary if the forms are to persist as separate species. First, they must be sufficiently different genetically not to interbreed freely, and secondly, they must be sufficiently different in their ecology to avoid serious competition with each other.

Chapter XVI: ADAPTIVE RADIATION

Natural selection will always act according to the nature of the places which are either unoccupied or not perfectly occupied by other beings; and this will depend on infinitely complex relations. But as a general rule, the more diversified in structure the descendants from any one species can be rendered, the more places they will be enabled to seize on, and the more their modified progeny will be increased.

Charles Darwin: *The Origin of Species*, Ch. III

SPECIES-FORMATION AND ADAPTIVE EVOLUTION

Many of the differences between the subgenera and genera of Darwin's finches are undoubtedly adaptive. This applies particularly to their beaks, as described in Chapter VI, but also to numerous other characters and habits. Thus the ground-finches hop about the ground, the insectivorous tree-finches are agile and tit-like among the branches, the woodpecker-finch climbs vertical trunks and inserts a cactus spine into crevices, the Cocos-finch and the cactus ground-finch have bifid tongues, and *Certhidea* has the quick flitting movements of a warbler. Like a warbler, too, *Certhidea* repeatedly flicks the wings partly open when hopping about the bushes. The reason for this habit is not known, but it is found in no other of Darwin's finches, and its parallel evolution in *Certhidea* and in the true warblers presumably means that it has some significance in the lives of these birds.

Darwin's finches provide an example of an adaptive radiation, with seed-eaters, fruit-eaters, cactus-feeders, wood-borers and eaters of small insects. Some feed on the ground, others in the trees. Originally finch-like, they have become like tits, like woodpeckers and like warblers. There are, of course, gaps; for instance, none have taken to open country, none have become birds of prey, and none are aquatic. But the considerable similarity between all the species in regard to breeding habits, plumage and internal structure implies that their divergence has been comparatively recent and rapid. Though scarcely in the same class with the astounding radiation of the Australian pouched mammals, Darwin's finches are sufficiently impressive.

Adaptive evolution, as now generally agreed, is brought about by the action of natural selection on random mutations. Those mutations which happen to be favourable spread through the population, and are repeatedly modified by further mutations, thus gradually building up the finely adapted structures and behaviour patterns which make up each living organism. This process is long-term and accumulative. At first sight it has nothing to do with the origin of new species, which is essentially the formation of gaps between populations. For this reason, and influenced by the apparent absence of adaptive differences between closely related species, some recent writers have claimed that the origin of species and long-term adaptive evolution are independent processes, and that the former has no important influence on the latter. This view I believe to be mistaken.

The point is illustrated by the sharp-beaked ground-finch *Geospiza difficilis* in the Galapagos. Probably in former times this bird bred in both the arid and the humid forest of the central islands, but it has since been driven from all except the humid zone by a newer species, the small ground-finch *G. fuliginosa* (see p. 28). Hence on the central islands *G. difficilis* can now become increasingly specialized for life in the humid forest. But this would not have been possible for it so long as it also bred in the arid zone, and is therefore a direct result of the appearance of the newer species *G. fuliginosa*.

Similar considerations apply wherever two species, originally geographical races of the same species, have become established in the same region. In most cases adaptive or ecological differences probably arose between the two forms during their period of geographical isolation before they met, but their meeting must almost inevitably have resulted in further restriction of their foods or habitats, and so must have tended to accelerate their adaptive specialization. For instance, the blue chaffinch of the Canary Islands probably occurred formerly in both pine and broad-leaved forest, but since the arrival of the second chaffinch species it has been confined to pine forest, after which, but not before, it could become specialized for the latter habitat.

Likewise the earlier and large-beaked species of the white-eye *Zosterops* on Lord Howe Island probably had a more varied diet before the arrival of the second and smaller species. The meeting of these two species in the same area may be expected to have led to a restriction of their diets, and thereafter to an increase in the size-difference between them.

The origin of new species, so far from being irrelevant, is therefore an active agent in promoting specialization, and in particular it is an essential precursor to an adaptive radiation. I consider that the adaptive radiation of Darwin's finches can have come about only through the repeated differentiation of geographical forms, which later met and became established in the same region, that this in turn led to subdivision of the food supply and habitats, and then to an increased restriction in ecology and specialization in structure of each form. On Cocos, where conditions are unsuitable for species-formation, there has likewise been no adaptive radiation among the land birds.

Various investigators have sought to explain the evolution of Darwin's finches through unusual genetic factors, such as an excessive inherent variability or frequent hybridization. But it is the persistence, rather than the origin, of new types which chiefly requires explanation, and in my opinion the peculiarities of the finches are primarily due to an unusual combination of geographical and ecological factors. The chief geographical factor has been the existence of a number of islands, which has provided favourable opportunities for the differentiation of forms in isolation and their subsequent meeting. The chief ecological factor has been the scarcity of other passerine birds, which has permitted an unusually great degree of ecological divergence between forms, and thus has allowed an unusually large number of related species to persist alongside each other without competition. Divergence may have been accelerated by the scarcity of predators (see p. 114), but this would seem, at most, very subsidiary

NATURE OF THE GENUS IN SONG-BIRDS

The genus is essentially an artificial unit for the convenient cataloguing of similar species. However, in continental passerine birds, systemists have tended to attach particular importance to morphological divergence, including beak differences, as a criterion for generic separation. On the continents, closely related passerine species do not usually differ greatly from each other in beak or other morphological characters. Hence the use of morphological characters for separating genera has tended to give a convenient number of species in each genus, and so has proved useful in practice.

On oceanic islands, a greater proportion of closely related passerine species differ prominently from each other in beak than is the case on the continents. This is not because beak differences are more frequent among insular than continental land birds—the reverse is usually the case—but because on oceanic islands the species which differ in beak are often very similar in other respects, and so tend to be placed in the same genus, whereas on the continents they are usually distinctive in other ways as well, and so tend to be placed in separate genera. Thus in Britain the small-beaked goldfinch, the medium-sized greenfinch, and the large-beaked hawfinch are related species living in similar habitats but eating mainly different foods. They therefore provide a close ecological parallel with the small, medium and large ground-finches of the Galapagos. But whereas the latter species differ in little except beak and are placed in the same genus *Geospiza*, the three British finches differ also in plumage, nesting habits and various other ways, and are placed in separate genera, *Carduelis*, *Chloris* and *Coccothraustes* respectively.

(i) (ii) (iii)

Fig. 25. Heads of (i) hawfinch, (ii) greenfinch and (iii) goldfinch. $\frac{2}{3}$ natural size (*after* ten Kate).

In applying generic names to oceanic land birds, the taxonomist is often in somewhat of a dilemma. Related species frequently differ so much in beak that, were they mainland birds, they would unhesitatingly be placed in different genera. But in other respects, including plumage, they are often very similar, indicating close relationship. If every island land bird that differs markedly in beak is placed in a separate genus, many of the resulting genera contain only a single species, which not only multiplies names excessively, but obscures the close relationship of forms that differ in little except beak. The difficulty discussed in Chapter II regarding the number of genera of Darwin's finches is not merely a question of terminology, but reflects the fact that on oceanic islands the land birds are at an earlier stage of evolution than on the continents. On the continents the divergence into finches, warblers, tits and woodpeckers obviously took place in a much more distant past than the corresponding divergence among Darwin's finches in the Galapagos.

BIRDS ON OTHER OCEANIC ISLANDS

An adaptive radiation like that of Darwin's finches is rarely found on other oceanic islands. For reasons already considered it was not to be expected on solitary islands, however remote. Such islands often have a single highly peculiar species, but in no case is there a group of related birds. The greatest number known is on Norfolk Island, 800 miles from Australia, where there are three species of white-eye *Zosterops*, differing conspicuously in size and in size of beak (Mathews, 1928; Stresemann, 1931). But these species have almost certainly evolved as a result of three separate colonizations of Norfolk Island from outside, either from the Australian mainland or from Lord Howe Island. There is no reason to think that they became differentiated on Norfolk Island itself, which has an area of less than 20 square miles.

It is more surprising that no parallel with Darwin's finches is found on most oceanic archipelagos. For example, the Azores are volcanic islands and further from a continent than are the Galapagos. But their song-bird inhabitants differ only subspecifically from those of Europe, and no endemic form has given rise to an adaptive radiation. The Azores have been colonized by a larger number of passerine birds than have the Galapagos, which suggests that they are more accessible. This is corroborated by the frequent records of European song-birds passing through the Azores on migration, whereas only two species of mainland passerines have been recorded on migration in the Galapagos, the American barn swallow *Hirundo erythrogaster* and the bobolink *Dolichonyx oryzivorus* (Swarth, 1931). This may be correlated with the fact that the Azores are in an area of winds, whereas the Galapagos are in a region of calms. Though in the Galapagos the south-east trade wind blows for part of the year, it does not blow from the South American mainland but forms at sea, away from the land.

The land organisms of remote islands seem less efficient than continental types, perhaps because, owing to the greater number of both individuals and species on the continents, competition is more severe there. When a remote island receives new bird colonists from a mainland area, they often eliminate the original inhabitants. This is being demonstrated in New Zealand at the present time, where artificially introduced European birds are spreading rapidly at the expense of the native species. Similarly, Usinger (1941) reports that in the Hawaiian Islands many of the peculiar endemic insects are being eliminated by introduced continental types. This suggests that even if the Azores were at one time so isolated that an endemic group comparable with Darwin's finches evolved there, such peculiar forms would probably have been eliminated later, with the arrival of the newer and more efficient species which now frequent the islands. That the latter birds are only recent arrivals from the mainland is shown by the slight extent to which they differ from European forms.

Similar considerations probably account for the absence of local adaptive radiations among the land birds of the Canary Islands, the Cape Verde Islands, the Revilla Gigedos, and other archipelagos off Europe, Africa and America. Even the Polynesian archipelagos provide no parallel with Darwin's finches. They possess a variety of land birds, but these are derived from the lands to the west or from adjoining archipelagos. No Polynesian bird has diverged into a group of species within one archipelago, presumably because invasions of new and more efficient species from outside have occurred at too frequent intervals. Though the distances between some of the Polynesian archipelagos are considerable, all are in a region of winds.

There are in the world only two archipelagos whose land birds show an adaptive radiation comparable with that of Darwin's finches. The three islands of the Tristan da Cunha group lie in the Atlantic some 2000 miles from South America and nearly as far from Africa. Aided by strong westerly winds, two South American passerine birds have colonized the islands, a thrush *Nesocichla* and a finch *Nesospiza*. The thrush is a single species and need not be considered further. But the finch has evolved into two species, one large and large-beaked, the other small and small-beaked, as shown in Fig. 24 (p. 139). Both species occur together on Nightingale Island, where Hagen (1940) found that they differ in food and habitat. The smaller species is divided into three island races, differing in size of beak, but the Tristan form has become extinct. Two hundred miles away from Tristan lies Gough Island, and here occurs a related finch *Rowettia* (*Nesospiza*) *goughensis*, rather similar to *Nesospiza* in appearance but with a more elongated beak. This bird may have evolved from *Nesospiza* stock, though Lowe considers that it was independently derived from South America (Clarke, 1905; Lowe, 1923).

As Lowe points out, Tristan da Cunha is Galapagos in miniature. At the other extreme are the astonishing Drepanididae, the sicklebills of Hawaii. The Hawaiian Islands are as remote in the Pacific as is Tristan da Cunha in the Atlantic, and only five passerine forms have succeeded in reaching them. Of these, the crow belongs to the mainland genus *Corvus*. The flycatcher *Chasiempsis* is placed in a distinct genus, and there is one form on each island. The thrush *Phaeornis* is also peculiar to Hawaii; there is one form on each island, while on Kauai occur two species which differ in size. The fourth colonist, a honeyeater, has diverged into two distinct genera, *Chaetoptila* and *Moho*. Finally, the fifth colonist, the sicklebill, has produced a multitude of forms far more diverse than Darwin's finches.

The ancestor of the sicklebills is considered to have been a finch, but it has given rise to some strikingly unfinchlike forms. A modern study of these birds has not been undertaken, but Rothschild (1893–1900) and Perkins (1903) divide them into as many as eighteen genera, some of which would probably be merged on modern standards. In most cases the genera are represented by distinctive island forms, formerly classified as

Fig. 26. Adaptive radiation of Hawaiian sicklebills (*after* Keulemanns).

separate species, while in *Chlorodrepanis*, also in *Rhodocanthis*, two related species differing prominently in size are found on the same island. Some genera feed on both nectar and insects, some mainly on nectar, and others mainly on insects. Others eat fruit, others seeds, one feeds mainly on a native bean and on caterpillars, and another on nuts. Some of the associated modifications in beak are shown in Fig. 26. The most remarkable is in *Heterorhynchus*, This bird climbs trunks and branches like a woodpecker, and feeds on longicorn beetle larvae in the wood. It taps with its short lower mandible, and probes out the insects with its long decurved upper mandible. This method may be compared first with that of a true woodpecker, which excavates with its beak and probes with its long tongue; secondly with that of the Galapagos woodpecker-finch *Camarhynchus pallidus*, which trenches with its beak and probes with a cactus spine; and thirdly with that of the extinct New Zealand huia *Heterolocha acutirostris*, in which the male excavated with its short beak, the female probed with its long decurved beak, the pair to some extent co-operating (Buller, 1888).

OTHER ORGANISMS ON ISLANDS

It is significant that the regions which have given rise to adaptive radiations among birds have produced a similar multiplicity of related species in various other land organisms. The most striking case in the Galapagos is that of the land mollusc *Nesiotus*, which has evolved into fifty-three species, or sixty-six geographical races (Dall and Oschner, 1928*b*; Gulick, 1932). A much smaller radiation has occurred in the iguanas, of which the two Galapagos genera probably evolved within the archipelago from one original colonist. The land iguana *Conolophus* lives on land and feeds primarily on cactus, while the marine iguana *Amblyrhynchus* lives on the shore and feeds primarily on marine green algae. Both iguanas show some differentiation into island forms.

In the other Galapagos reptiles, evolution has proceeded only to the stage of differentiation into well-marked geographical forms. The giant tortoise *Testudo* is divided into as many as fifteen geographical forms, five of which occur on separate parts of Albemarle, and there are seven or eight island forms of the lizard *Tropidurus*, the gecko *Phyllodactylus* and the snake

Dromicus[1] (Van Denburgh, 1912–14; Van Denburgh and Slevin, 1913).

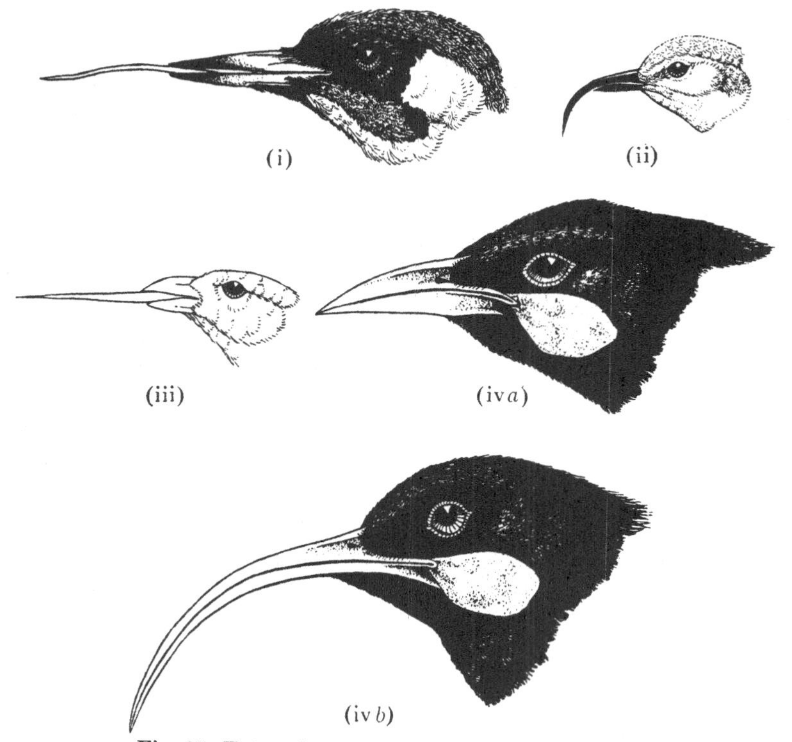

Fig. 27. Extraction of insects from wood by birds.

(i) European green woodpecker *Picus viridis* excavates with beak, probes with long tongue.

(ii) Hawaiian *Heterorhynchus* taps with short lower mandible, probes with long upper mandible. (Head *after* Keulemanns.)

(iii) Galapagos woodpecker-finch *Camarhynchus pallidus* trenches with beak, probes with cactus spine.

(iv) New Zealand huia *Heterolocha acutirostris*: (*a*) male excavated with short beak, (*b*) female probed with long beak. (Heads *after* Keulemanns.)

Note. Various other trunk-feeding birds either excavate only or probe only, but they do not do both.

Heads drawn ½ natural size.

Well-marked island forms, from four to nine in number, are also found in the Galapagos ants *Camponotus macilentus* and

[1] Actually two species of the snake *Dromicus* are found on Indefatigable and Jervis. Van Denburgh considers that these did not evolve from each other within the Galapagos, but colonized the islands independently from South America.

C. planus (Wheeler, 1919), and in the four grasshoppers *Halmenus*, *Liparoscelis*, *Schistocerca* and *Sphingonotus* (McNeill, 1901; Snodgrass, 1902*b*; Hebard, 1920). It is curious that no Galapagos insect is recorded as having produced an adaptive radiation, with a multiplicity of species living on each island. Such radiations are common on other oceanic islands, so further collecting may possibly reveal their existence in the Galapagos.

One of the Galapagos plants has produced a radiation, the endemic helianthoid genus *Scalesia* having evolved into a group of trees and bushes consisting of six main species, each of which is in turn divided into a number of geographical forms (Howell, 1941). Groups of up to about a dozen endemic Galapagos species are also found in the shrubs and bushes of the genera *Telanthera* (Amarantaceae), *Acalypha* and *Euphorbia* (Euphorbiaceae), *Cordia* (Boraginaceae) and a few others. These genera are not endemic to the Galapagos, and botanists do not state whether the various species are the result of independent colonization from outside regions, or whether, as in *Scalesia*, a group of species has evolved within the archipelago from one original form. Some of the species are divided into island forms, a good example being *Euphorbia viminea*, with eight geographical forms (Robinson, 1902; Stewart, 1911).

In other groups of land organisms, as in birds, the most remarkable radiations are those which have occurred in the Hawaiian Islands. The most remarkable of all is that of the achatinellid land molluscs, of which there are nearly five hundred known forms, all apparently descended from one original form on the islands (Usinger, 1941).

The many striking radiations among Hawaiian insects have been described by Perkins (1913). They include three endemic subfamilies and four other genera of Coleoptera, five endemic genera of Lepidoptera, and from one to three endemic genera of Orthoptera, Odonata, Hemiptera, Neuroptera and Hymenoptera. Each of these groups presumably evolved from one original species, though each now contains at least thirty species, while in the *Proterhinus* weevils, the Anchomenini ground-beetles and the *Hyposmochoma* moths, there are well over a hundred related species in each group. These are the figures given by Perkins, and since he wrote, the number of known forms has been considerably increased. In numerous cases an insect is represented by a different form on different islands, while when closely related species are found on the same island, they often occur on different food plants. Perkins concluded that geographical isolation has been the factor of primary importance in the origin of the insect species.

Radiations have also occurred among Hawaiian plants. Thus according to Gulick (1932), there are 146 species and varieties of Hawaiian lobelias, these having evolved from an extremely small number of original colonists. The survey by Hillebrand (1888) shows that there are several other genera of trees and shrubs endemic to Hawaii and represented in the islands to-day by between ten and thirty species, presumably as a result of local evolution within the archipelago.

In Polynesia, as already mentioned, no local radiations are found among the land birds, but they occur in some cases in the trees, insects and land molluscs (Buxton, 1938; Crampton, quoted by Huxley, 1942, etc.). In general, the insects, land molluscs and land plants have produced more local radiations on remote islands than have the land birds. It is of special interest to find such radiations on the solitary island of St Helena, which is in the Atlantic 1200 miles from West Africa and remote from all other islands. Wollaston (1877) described as many as ninety-one species of weevils endemic to St Helena, and five of the endemic genera have each given rise to between ten and fourteen species, presumably as a result of local evolution on St Helena. One endemic genus of land molluscs has also diverged into about a dozen species (Smith, 1892), and there seems to have been a small radiation among trees of the family Compositae (Melliss, 1875).

No single island, however remote, has given rise to an adaptive radiation of land birds. This is probably because an essential step in the formation of new bird species is the isolation of geographical forms. For the latter, a single small island does not suffice, and either an extensive land area or a number of small islands is required. That the insects and land molluscs have produced radiations on a solitary island, such as St Helena, suggests that populations of these organisms can be isolated topographically in a much smaller space than bird populations. This is corroborated by the work of Crampton, Gulick and others, who have shown that in some of the Hawaiian and Polynesian land molluscs, each steep valley may have its own subspecies.

The comparative ease with which insects and land molluscs are isolated means that in these groups new species tend to arise more quickly than they do in birds. This is presumably one reason why in Hawaii and elsewhere the radiations among insects and land molluscs are more extensive than those of birds. But another, and probably more important, reason is that a much greater number of ecological niches are available for insects and land molluscs than for land birds. It is not only the origin, but also the persistence, of new species which requires explanation.

That on oceanic islands radiations have occurred not only among the land birds, but also among the insects, land molluscs and land plants, suggests that the evolutionary factors involved in these other groups are fundamentally the same as they are in Darwin's finches. This is one of the chief reasons why I hope this book may have been worth writing.

L'ENVOI

The drawers of the Rothschild collection contain more representatives of some of the Hawaiian sicklebills than are alive in the islands to-day. These remarkable birds could not survive the changes in the vegetation and animal life caused by the civilization of the white man. The finches on Tristan and the indigenous white-eyes on Lord Howe have also gone. Much the same has happened or is happening on all other oceanic islands, and indeed everywhere in the world. Through carelessness or for commercial profit, we are allowing the natural cover and adornment of the earth to disappear.

By a fortunate chance Fray Tomàs de Berlanga had to allay his thirst by chewing cactus, and neither he, nor the crew of the *Bachelors' Delight*, nor others for more than a century longer, seem to have appreciated that human settlement might be possible in the Galapagos. As a result, permanent occupation was delayed until well on into the nineteenth century. Then, of course, havoc followed fast, but as yet few of the small land birds have been affected, so that by a miracle Darwin's finches have survived into our own time.

Among continental birds the main adaptive radiation took place too far back for its origins to be at all clear. On the other hand, the situation on most oceanic islands is too simplified to throw light on the main processes of evolution, though the *Zosterops* on Norfolk and Lord Howe and the *Nesospiza* on Tristan de Cunha provide valuable pointers. But the evolutionary stage shown by Darwin's finches is both sufficiently advanced to provide a parallel with the more mature evolution of the continents, and sufficiently early for links to remain which reveal the underlying processes.

It may be supposed that the ancestral finch first became differentiated into various forms in geographical isolation on different islands. After a sufficiently long period of isolation, some of these forms had become so different that, when by chance they met on the same island, either they were already intersterile, or hybrids between them were at a selective disadvantage so that intersterility was evolved. Thus new species originated. But when they met, such new species must have tended to compete with each other; both could persist together only if they had diverged sufficiently for each to prove better adapted to one part of the original food supply or habitat, which was then divided between them. After such restriction in ecology, each became increasingly specialized. Each then spread to other islands, so that it became differentiated into new geographical forms, some of which in their turn later met on the same island and kept distinct, thus giving rise to further new species, and resulting in further ecological restriction and structural specialization. In this way there gradually appeared the adaptive radiation whose end-forms are alive to-day.

The evolutionary picture presented by Darwin's finches is unusual in some of its details, but fundamentally it is typical of that which I believe to have taken place in other birds, while many of the same general principles probably apply to other groups of animals and to plants. Darwin's finches form a little world of their own, not, however, a peculiar world, but one which intimately reflects the world as a whole, so that with these birds, as Darwin wrote, we are brought somewhat nearer than usual 'to that great fact—that mystery of mysteries—the first appearance of new beings on this earth'.

SUMMARY

Part I: Description

I. A brief and depressing description is given of the Galapagos islands, their vegetation and their human history, including the main ornithological visits.

II. The principles adopted in naming birds are discussed, with some peculiar difficulties found in Darwin's finches. Tables are given of the genera, species and races (subspecies) of Darwin's finches and their distributions on the various islands. Each Galapagos island supports between three and ten species and there are fourteen species in all, one of which lives on Cocos. Some inter-island wandering has been recorded. Two of the forms collected by Darwin have since become extinct.

III. In two instances, two related species occupy separate but adjacent habitats, in one case clearly due to interspecific competition. The factors limiting habitat distribution are discussed generally. Most of Darwin's finches are not isolated from each other either in habitat, nesting site or breeding season, and they freely use each others' nests when displaying. The birds have scarcely any natural enemies and are extremely tame.

IV. The females of all the species are dully coloured. Small differences in plumage are found both between some of the species and between some of the races of the same species. Some of these differences seem adaptive, but others appear to be without adaptive significance.

V. The display of all the species is the same, and their songs are also very similar. In some species the fully plumaged male is wholly black, in others partly black, and in others without black, but in the first two cases many individual males breed without black or in an incompletely black condition. There is a general tendency for the land birds of remote islands to lose distinctive male plumage, probably because they have normally got away from all similar species, so that the need for specific recognition marks disappears. The ancestral Darwin's finch has subsequently diverged into a group of species, which recognize each other primarily by beak differences.

VI. The marked beak differences between the subgenera of Darwin's finches are correlated with marked differences in their diets and feeding methods. On the other hand, closely related species have similar feeding methods; they differ markedly in the size of their beaks, which is attributed to their taking foods of the same general nature but of different size. On outlying Galapagos islands a species sometimes occupies an unusual food niche, or takes foods which on the central islands are divided between two species, and in these cases the beak of the island form is correspondingly modified.

VII. In many cases there are average differences in size of beak and in wing-length between island forms of the same species. Some of these differences are adaptive, others seem without adaptive significance. In some cases island forms differ in average size, though occupying the same ecological niche on islands only a few miles apart.

VIII. Three species of *Geospiza* differ from each other solely in absolute size and in proportions, and the extreme individuals of one species come very close to, or actually overlap with, the extreme individuals of another. Similar relations hold for three species of *Camarhynchus*.

IX. Some forms of Darwin's finches are more variable than others, and four species are exceptionally variable. The factors influencing variability are discussed.

X. Intermediate and freak specimens are unusually common in Darwin's finches, and they include two which are apparently intergeneric hybrids. Several species and island forms are intermediate in appearance between other forms, but this is probably due to their occupying an intermediate food niche, and not to a hybrid origin. There is no reason to think that hybridization has played an important part in the evolution of Darwin's finches.

XI. All the species are closely related, and suggestions are made as to the way in which they may have evolved from each other.

PART II: INTERPRETATION

XII. There is no reason to think that the Galapagos were formerly connected by land with the American mainland, and the Galapagos animals have presumably arrived by occasional means over the sea. Only six passerine forms and one cuckoo have colonized the islands, and these show a very varying degree of divergence from their mainland ancestors. Darwin's finches probably arrived before any other land birds, which has permitted their evolution in unfinchlike ecological niches. They have been comparatively free from predators.

XIII. Island forms of the same species show every gradation from extremely small differences in average size to differences as great as those which separate some of the species. The differences tend to be greater the greater the degree of isolation. Some of them are adaptive and others seem non-adaptive.

XIV. New species originate when forms differentiated in geographical isolation later meet in the same region and keep distinct. A good example occurs in *Camarhynchus* on Charles, and cases are also given in birds from other parts of the world. There is not as yet any evidence that new species of birds arise in other ways. The evolution of specific intersterility is discussed.

XV. When two related bird species meet in the same region, they tend to compete, and both can persist there only if they are isolated ecologically either by habitat or food. All of Darwin's finches are isolated from each other ecologically. Ecological isolation is discussed in other birds, and in other types of organisms. Closely related bird species differ from each other adaptively, though such differences are often inconspicuous. Adaptive and ecological divergence probably arises at the subspecific level, though intensified after the forms have met in the same area.

XVI. The meeting of two forms in the same region to form new species must, when both persist, result in subdivision of the food or habitat, and so to increased specialization. The repetition of this process has produced the adaptive radiation of Darwin's finches. A radiation similar to that of Darwin's finches is not found in the land birds of single oceanic islands, because these provide no opportunities for the formation of new species in geographical isolation. Instances are rare even on other archipelagos, probably because most archipelagos are too accessible to fresh colonization by more efficient birds from outside areas. Two parallel radiations are known, the finches of Tristan da Cunha being at a much earlier stage, and the sicklebills of Hawaii at a much more advanced stage, than Darwin's finches. Radiations are also found in the land insects, land molluscs and land plants of remote islands, including single islands. The manner of evolution of Darwin's finches is peculiar in some details, but is considered to be fundamentally typical of that which has occurred in many other organisms.

ACKNOWLEDGEMENTS

The expedition

The expedition to the Galapagos would not have been possible for me without substantial grants from the Royal Society and the Zoological Society of London; further financial assistance was received from the Elmgrant Trustees, and another member of the party, L. S. V. Venables, was helped by grants from the British Ornithologists' Union, the British Ornithologists' Club and certain private ornithologists. To all of these grateful acknowledgment is made.

The expedition would also not have been possible without its other members, W. H. Thompson and L. S. V. Venables, ornithologists, R. Leacock, photographer, and Mr and Mrs T. W. J. Taylor, botanists, to all of whom most grateful thanks are due. In addition to their own studies, which are appearing elsewhere, each of the above made some contribution to the work described in this book. W. H. Thompson carried out extensive and valuable field observations on the finches, particularly on *Certhidea*, which is the most difficult species to watch, and by his skill and experience he greatly clarified the investigation of the breeding cycle in these birds. L. S. V. Venables undertook much of the construction of the aviaries, and all the subsequent care of the birds, in the thankless task of attempting to get finches of different species to interbreed in the Galapagos. R. Leacock took most of the photographs used in this book, some of which are from the 16 mm. film of the Galapagos which he photographed, and which was edited by the late W. H. Hunter and myself and issued by the Dartington Hall Film Unit, Totnes. The other photographs used in this book were taken by T. W. J. Taylor, who together with Mrs Taylor instructed me in the botany of the islands.

A very considerable debt is owing to Dr Julian S. Huxley, who took an extremely stimulating and practical interest in the expedition from start to finish, and helped in innumerable ways, while his views, both in discussion and in published works (Huxley, 1940, 1942), were of great value when I came to write up results. I must also thank Dr P. R. Lowe, since, though I have disagreed with some of his conclusions, it was his stimulating paper (Lowe, 1936) which first suggested the idea of studying the finches in the Galapagos.

The Ecuadorean government kindly accorded permission for the expedition to stay in the islands. The British consul and vice-consul in Guayaquil assisted with customs, permits and other formalities, and helped to arrange our passage. Various residents in the Galapagos were of considerable assistance, in particular the Cobos family in whose house we stayed on Chatham, and the Küblers and Angermeyers on Indefatigable. It is also a pleasure to record the help of various firms and organizations who gave us goods either free or at reduced prices: Australian Dried Fruits Board, Buck and Hickman Ltd., Burroughs Wellcome and Co., Camp and Sports Co-operators Ltd., Chivers and Sons Ltd., T. Cook and Sons Ltd., Frank Cooper Ltd., Grape Nuts Co. Ltd., C. T. Harris (Calne) Ltd., Huntley and Palmer Ltd., Kearley and Tonge Ltd., Robert Lawrie, Lever Bros., Port Sunlight Ltd., Nestlé's Milk Products Ltd., James Robertson and Sons Ltd Ryvita and Co. Ltd., Spillers Ltd., and Wisbech Produce Canners Ltd. The goods from these sources did much to alleviate the rather trying conditions of camp life in the Galapagos, and the foods in particular meant more to us than the above list might suggest to those accustomed to the comparatively well-stocked larders of war-time England.

The captive finches

It was originally intended to bring captive individuals of Darwin's finches back to England to determine whether different species could be induced to interbreed. Unfortunately, when we left the Galapagos our captive finches were in full moult, and by the time that they reached Panama they were in such bad condition that it seemed very doubtful whether they would reach England alive. Accordingly, we cabled Dr Julian S. Huxley, Secretary of the London Zoological Society, and Dr Robert C. Miller, Director of the California Academy of Sciences, San Francisco, and with the consent of both parties, the finches were taken the much easier journey to California. We selected the California Academy of Sciences for this purpose because this institution has long made a special study of the Galapagos fauna and flora, and particular thanks are due to them for undertaking the responsibility of keeping the birds at such short notice.

On arrival in San Francisco, the birds were taken to Mr E. C. Kinsey, an amateur aviculturist of great experience, and through his care and skill, nearly all the birds were saved and brought back into good condition. The birds were housed in Mr Kinsey's private aviaries until special buildings had been erected for them at the California Academy of Sciences. At the latter, individuals of several species have bred successfully with others of their own kind, though not with members of different species, and Dr Robert T. Orr is preparing an account of his observations on the birds in captivity. The outbreak of hostilities in Europe while I was still in the United States made it inadvisable to bring any of the captive birds back to England at that time, but it is hoped to do this eventually.

Work in museums

While at the California Academy of Sciences, I took the opportunity to make a detailed study of the extensive collections of Darwin's finches made by the Academy expedition of 1905–6. This large series of birds made possible a statistical treatment, which was considerably assisted by work on the large collections at the American Museum of Natural History, New York, at the U.S. National Museum at Washington, at Stanford University, at the Museum of Comparative Zoology, Cambridge, Mass., and at the British Museum of Natural History, while the small series at the Museum of Vertebrate Zoology, Berkeley, the Carnegie Museum, Pittsburgh, the Field Museum, Chicago, and the University of Michigan were also examined. To the staffs of all these institutions very grateful thanks are due for providing me with every facility for work, and in many cases they also provided stimulating discussions and extended hospitality. I am particularly indebted to Dr Robert C. Miller, the late J. Moffitt, Dr R. T. Orr and others at the California Academy of Sciences, to Dr Ernst Mayr in New York, Dr Alden H. Miller at Berkeley, and Mr N. B. Kinnear at the British Museum. I am also greatly indebted to Dr R. A. Fisher, for his advice on the statistical side of the work.

THE BOOK

The whole or part of the draft manuscript of this book was read and carefully criticized by E. A. Armstrong, A. C. Fabergé, S. Smith, H. N. Southern, W. H. Thorpe, and G. C. Varley, and their comments were of considerable value in preparing the final revision. My views on interspecific competition were largely formulated in discussions with G. C. Varley, to whom I am particularly grateful. I must also thank E. Mayr, R. E. Moreau and B. B. Roberts for valuable comments on the book in proof.

The coloured plates are reproduced from illustrations by John Gould in the *Zoology of the Voyage of the H.M.S. 'Beagle'* (1841). The drawing of the fourteen species of finches (Fig. 3) is by Lt.-Col. W. P. C. Tenison, D.S.O.

The sketches of *Camarhynchus pallidus* with its stick (Fig. 6) are by Roland Green from photographs by R. Leacock. Fig. 27 is by C. E. Talbot Kelly, the heads of the huia and of *Heterorhynchus* being after J. G. Keulemanns in Buller (1888) and Rothschild (1893–1900) respectively. The other text drawings of birds are by Susan Williams-Ellis, her sources being as follows: for the heads and beaks of Darwin's finches, the figures by F. Abernathy in the monograph by Swarth (1931); for the heads of *Nesospiza,* figures by H. Grönvold in Lowe (1923); for the pictures of *Zosterops* on Lord Howe, the illustrations by H. Grönvold in Mathews (1928); for the heads of goldfinch, greenfinch and hawfinch the illustrations in ten Kate (1937); and for the adaptive radiation of the Hawaiian sicklebills, the illustrations by J. G. Keulemanns in Rothschild (1893–1900). The map diagrams were prepared by Madge Humphreys. I am greatly indebted to all these illustrators for the care and skill which they have expended on their work.

As already mentioned, the photographs of the Galapagos scenery and animals were taken by R. Leacock and T. W. J. Taylor, and I am also indebted to the Dartington Hall Film Unit for obtaining prints from the negative of the 16 mm. film on the Galapagos. The heads of the *Geospiza* specimens in Plate VII were photographed by the staff of the California Academy of Sciences from specimens in their collection.

TABLES OF MEASUREMENTS

(for male individuals only, unless specified to the contrary)

TABLE XVIII (see text, pp. 48–49). PERCENTAGE OF COLLECTED MALES IN PARTLY BLACK PLUMAGE IN *CAMARHYNCHUS*

Island	*C. crassirostris*	*C. psittacula*	*C. pauper*	*C. parvulus*
Abingdon	(50)	(50)	—	—
Bindloe	—	54	—	—
James	77	(82)	—	64
Indefatigable	57	—	—	(26)
Albemarle	60	(48)	—	58
Chatham	24	—	—	4
Charles	24	—	26	70

Notes. (i) Brackets indicate only 16–19 specimens available; otherwise at least 24. For number of specimens available, see Supplementary Table A (p. 183).

(ii) The collected sample may not truly reflect the real percentage on each island. First, collectors tend to select the best-plumaged males. To set against this, most collecting has been done near the coast, but on the larger islands the fully plumaged males tend to breed inland. This is because the fully plumaged males tend to breed earlier than the others, and the transitional forest becomes suitable for breeding earlier than the lowland zone. For the latter reason the proportion of each type collected also depends on the season when the collection was made.

TABLE XIX (see text, pp. 48–49). PERCENTAGE OF COLLECTED MALES IN FULLY BLACK PLUMAGE IN *GEOSPIZA*

	Geospiza					
Island	*magnirostris*	*fortis*	*fuliginosa*	*difficilis*	*scandens*	*conirostris*
Culpepper	—	—	—	52	—	—
Wenman	—	—	—	39	—	—
Tower	(69)	—	—	83	—	72
Abingdon	35	25	27	(58)	—	—
Bindloe	34	39	(46)	—	—	—
James	35	56	(83)	75	(64)	—
Jervis	63	—	—	—	—	—
Indefatigable	—	29	41	70	49	—
Duncan	—	33	55	—	—	—
Albemarle	—	42	64	—	—	—
Barrington	—	—	66	—	64	—
Chatham	—	51	50	—	—	—
Hood and Gardner	—	—	73	—	—	64
Charles	—	58	47	—	57	—

Notes. (i) Percentages are based on at least 30 specimens except where placed in brackets, which denotes that between 21 and 29 are available. For number of males available, see Supplementary Table A (p. 183).

(ii) The number of males in partly black plumage varied from 7 to 35 per cent.

(iii) That small islands such as Tower and Jervis have a high proportion of black males may be simply because on large islands fully plumaged males tend to breed inland, and so farther away from the collector, whereas on small islands such males inevitably breed close to the coast (see also note (ii) to Table XVIII).

TABLE XX (see text, pp. 48, 73). RELATION OF WING-LENGTH TO BLACK MALE PLUMAGE

in one selected island population of each species

Species	Island	Average wing-length in mm. Males: Fully black	Males: Partly black	Males: Without black	Females
Geospiza					
magnirostris	James	84	81	81	80
fortis	Chatham	74	72	70	70
fuliginosa	Chatham	64	63	62	61
			Partly black and without black		
difficilis	Wenman	72	71		69
scandens	Indefatigable	73	71		69
conirostris	Gardner near Hood	79	77		75
Camarhynchus					
crassirostris	South Albemarle	—	84	83	81
psittacula	Bindloe	—	71	70	68
parvulus	Charles	—	64	63	61
Pinaroloxias					
inornata	Cocos	68	67	67	65
Camarhynchus			All males		
pallidus	South Albemarle		72		69
Certhidea					
olivacea	Indefatigable		54		53

Note. For number of specimens measured, see Supplementary Table B, p. 183.

TABLE XXI (see text, p. 73). RELATION OF CULMEN LENGTH TO BLACK MALE PLUMAGE

Species	Island	Average length of culmen (from nostril)		
		Black males[1]	Other males[1]	Females
Geospiza				
magnirostris	James	15·8	16·0	15·7
fortis	Chatham	12·3	12·0	12·0
fuliginosa	Chatham	9·0	8·4	8·6
difficilis	Wenman	10·7	10·7	10·6
scandens	Indefatigable	15·0	14·9	15·0
conirostris	Gardner near Hood	14·8	14·5	14·1
Camarhynchus				
crassirostris	South Albemarle	10·4	10·1	10·0
psittacula	Bindloe	10·6	10·4	10·3
parvulus	Charles	7·3	7·3	7·2
Pinaroloxias				
inornata	Cocos	10·5	10·3	10·3
Camarhynchus		All males		
pallidus	South Albemarle	11·2		11·0
Certhidea				
olivacea	Indefatigable	7·6		7·9

[1] Black males means fully black males in the case of all *Geospiza* spp. and *Pinaroloxias*, and males showing some black in *Camarhynchus* spp. Other males means males showing no black in *Geospiza magnirostris*, *G. fortis*, *G. fuliginosa*, *Camarhynchus* spp. and *Pinaroloxias*. But in *Geospiza difficilis*, *G. scandens* and *G. conirostris* the other males also include partly black males, as these are difficult to separate from those showing no black.

Only one form of each species has been tabulated. The others corroborate that the black males tend to have larger beaks than males without black. The slight difference is often not significant for one particular form, but is apparent in the series as a whole.

For number of specimens measured, see Supplementary Table B, p. 183.

TABLE XXII (see text, p. 74). SOME CORRELATION COEFFICIENTS

Species	Locality	Correlation coefficient of (i) wing: culmen	(ii) culmen: depth of beak
Geospiza			
magnirostris	(i) All, (ii) Abingdon	+0·42	+0·63
fortis	Charles	+0·67	+0·80
fuliginosa	Chatham	+0·25	+0·87
difficilis	James	(+0·11)	+0·52
scandens	Charles	(+0·21)	+0·43
conirostris	Hood	+0·49	+0·75
Camarhynchus			
crassirostris	South Albemarle	+0·48	+0·36
pauper	Charles	(+0·36)	+0·48
parvulus	Charles	(+0·13	+0·30
pallidus	South Albemarle	(+0·05)	(+0·09)
Certhidea			
olivacea	Indefatigable	(+0·30	—
Pinaroloxias			
inornata	Cocos	(−0·02)	—
Passer			
domesticus	Berkeley, California	None	+0·16
Junco			
19 forms	North America	Present in only 3 cases, values up to +0·55	Present in only 4 cases, values up to +0·60

Notes. (i) Brackets indicate correlation lacking or not significant (less than 20:1 probability). Numbers measured will be found under appropriate species and island in Supplementary Table C (p. 184). For wing: culmen correlation take second figure, and for culmen : depth correlation take first figure under each form. For *Passer domesticus* 87 were measured.

(ii) For wing : culmen correlation only fully black males were used in *Geospiza* spp. and *Pinaroloxias*, and only partly black males in *Camarhynchus* spp.

TABLE XXIII (see text, p. 74–79). MEAN SIZE DIFFERENCES IN DARWIN'S FINCHES

Measurements in mm. c = culmen, d = depth of beak, w = wing. Beak measurements for all males. Wing measurements for black males in *Geospiza* and *Pinaroloxias*, for partly black males in *Camarhynchus crassirostris*, *C. psittacula* and *C. parvulus*, for all males in other species.

	Geospiza														
	magnirostris			*fortis*			*fuliginosa*			*difficilis*			*scandens* and *conirostris*		
Island	*c*	*d*	*w*	*c*	*d*	*w*	*c*	*d*	*w*	*c*	*d*	*w*	*c*	*d*	*w*
Culpepper	—	—	—	—	—	—	—	—	—	11·3	9·0	73	15·0	16·5	82
Wenman	15·7	20·4	86	—	—	—	—	—	—	10·7	8·3	72	—	—	—
Tower	16·5	21·2	86	—	—	—	—	—	—	9·4	7·9	63	14·4	13·0	77
Abingdon	16·0	20·0	82	11·3	11·8	67	8·2	7·7	60	9·7	8·5	63	14·6	9·7	73
Bindloe	15·2	19·1	82	11·7	12·1	67	8·1	7·4	59	—	—	—	15·1	10·6	73
James	15·9	20·5	84	11·5	12·5	72	8·4	8·0	64	10·3	9·4	72	12·9	8·8	70
Jervis	15·3	18·6	83	11·5	12·6	—	8·5	8·2	64	—	—	—	13·6	9·2	71
Indefatigable	15·3	19·1	—	12·0	12·8	73	8·4	8·2	64	9·6	8·7	69	15·0	9·8	73
Duncan	—	—	—	11·3	11·5	70	8·6	8·1	64	—	—	—	—	—	—
North Albemarle	—	—	—	11·5	12·4	71	8·3	8·1	64	—	—	—	—	—	—
South Albemarle	15·3	18·4	—	12·5	13·9	75	8·4	8·3	65	—	—	—	14·6	9·6	—
Barrington	—	—	—	—	—	—	8·8	8·1	63	—	—	—	14·4	10·1	71
Chatham	—	—	—	12·2	13·2	74	8·8	8·1	64	—	—	—	13·3	9·7	—
Hood	—	—	—	—	—	—	8·8	8·3	64	—	—	—	15·4	16·0	80
Charles	—	—	—	11·7	12·5	72	8·6	8·1	64	—	—	—	13·9	9·4	71
Darwin's specimens	18·4	23·7	91	—	—	—	—	—	—	11·1	10·1	72	—	—	—

	Camarhynchus													*Certhidea olivacea*	
	crassirostris			*psittacula*			*parvulus*			*pallidus*				
	c	*d*	*w*	*c*	*d*	*w*	*c*	*d*	*w*	*c*	*d*	*w*	*c*	*w*
Culpepper	—	—	—	—	—	—	—	—	—	—	—	—	8·2	56
Wenman	—	—	—	—	—	—	—	—	—	—	—	—	7·8	55
Tower	—	—	—	—	—	—	—	—	—	—	—	—	8·2	54
Abingdon	10·4	12·2	85	10·1	10·4	71	—	—	—	—	—	—	8·1	52
Bindloe	—	—	—	10·5	10·5	71	—	—	—	—	—	—	8·2	54
James	10·6	12·8	87	9·8	11·2	74	7·0	7·4	64	12·6	9·3	75	7·3	54
Jervis	—	—	—	—	—	—	—	—	—	—	—	—	—	—
Indefatigable	10·4	12·3	86	9·6	10·7	—	7·5	7·3	64	12·1	9·1	73	7·6	54
Duncan	—	—	—	—	—	—	—	—	—	—	—	—	7·6	52
North Albemarle	—	—	—	—	—	—	7·0	7·4	63	11·7	8·8	73	7·3	53
South Albemarle	10·3	12·0	84	8·5	9·3	69	7·3	7·6	63	11·2	8·1	72	7·5	53
Barrington	—	—	—	—	—	—	—	—	—	—	—	—	7·9	53
Chatham	10·5	12·6	85	—	—	—	8·0	7·9	65	10·8	8·9	72	7·9	54
Hood	—	—	—	—	—	—	—	—	—	—	—	—	8·0	53
Charles	10·7	12·6	86	—	—	—	7·3	7·5	64	—	—	—	7·7	55

	c	*d*	*w*
Camarhynchus pauper Charles	9·0	8·8	70
C. heliobates Albemarle	10·5	8·2	72
Pinaroloxias inornata Cocos	10·4	6·2	68

Notes. (i) The depth of the beak is difficult to measure accurately in *Certhidea* as the beak is so narrow, being around 4·0 mm., but significantly thicker in birds from Charles.

(ii) For limits of size see Tables XX, XXI and XXII; for number of specimens measured see Supplementary Table C (p. 184); and for standard deviations, see Tables XXIX and XXX.

TABLE XXIV (see text, p. 82–84). LIMITS OF SIZE IN *GEOSPIZA* SPECIES

Island	(a) Length of culmen (males)			(b) Depth of beak (males)			(c) Wing length (black males)		
	G. fuliginosa	*G. fortis*	*G. magnirostris*	*G. fuliginosa*	*G. fortis*	*G. magnirostris*	*G. fuliginosa*	*G. fortis*	*G. magnirostris*
Wenman	Absent	Absent	14·8–17·0	Absent	Absent	19·0–21·5	Absent	Absent	84–88
Tower	Absent	Absent	15·3–17·7	Absent	Absent	18·6–23·2	Absent	Absent	84–89
Abingdon	7·5– 9·1	10·3–12·5	14·8–17·3	6·7–8·5	10·6–14·1	18·2–22·1	57–61	64–72	79–87
Bindloe	7·4– 8·7	10·6–12·5	13·8–17·4	6·7–8·2	11·4–12·9	17·1–21·1	56–61	65–70	77–85
James	7·4– 9·1	10·0–12·6	13·8–17·4	7·4–8·3	10·5–13·9	17·1–22·1	60–66	69–77	81–88
Jervis	7·9– 9·3	11·1–12·2	13·9–17·0	7·5–9·0	12·0–13·5	15·8–20·5	63–67	71–72	78–90
Indefatigable	7·5– 9·3	10·5–13·9	14·0–16·3	7·5–9·3	10·7–16·6	16·3–21·0	61–66	69–79	83–86
Duncan	7·7– 9·6	10·1–13·6	Present	7·3–9·0	10·0–14·8	Present	60–67	65–73	Present
North Albemarle	7·8– 9·2	10·1–13·3	14·4–16·0	7·5–9·0	10·3–15·6	Present	60–66	66–75	81
South Albemarle	7·9– 9·6	10·4–14·0	13·9–16·6	7·6–9·5	10·9–15·8	16·3–20·5	62–68	72–80	Present
Barrington	8·2– 9·6	Present	Present	6·8–8·8	Present	Present	61–66	Present	Present
Chatham	7·2–10·2	9·9–13·7	Absent	7·1–9·1	10·8–16·0	Absent	60–67	69–78	Absent
Charles	7·7– 9·4	9·8–14·2	(18·0–18·9)[1]	7·3–9·1	10·1–16·6	(23·6–23·7)[1]	61–70	67–80	(80–83)[1]
Hood	8·1– 9·7	Absent	Absent	7·3–9·4	Absent	Absent	61–67	Absent	Absent

Note. For number of specimens measured, see Supplementary Table C (p. 184).

[1] Darwin's extinct form.

TABLE XXV (see text, p. 88). LIMITS OF SIZE IN *CAMARHYNCHUS* SPECIES

Island	Culmen		Depth of beak		Wing (partly black males)	
	C. parvulus	*C. psittacula*	*C. parvulus*	*C. psittacula*	*C. parvulus*	*C. psittacula*
Abingdon	—	9·4–10·7	—	9·7–11·2	—	67–73
Bindloe	Absent	9·9–11·3	Absent	10·0–11·2	Absent	69–75
James	6·4–7·4	9·1–10·4	6·5–8·0	9·4–11·9	60–66	72–77
Indefatigable	6·6–8·2	8·8–10·3	6·9–8·1	9·8–11·3	—	—
South Albemarle	6·6–8·0	8·0– 9·5	6·9–8·1	8·5–10·3	62–65	67–72
Charles	6·8–8·1	8·0–10·2 }[1] 9·6–10·2	6·7–8·0	8·1–9·9 }[1] 11·1	62–68	68–74 }[1] —
Chatham	7·0–9·3	Absent	7·2–8·7	Absent	63–67	Absent

[1] Upper of two figures for Charles refers to *C. pauper* and lower to *C. psittacula.* Of the latter only 3 males are available, but the 10 females collected on Charles also showed no overlap with female *C. pauper* in depth of beak. For number of specimens measured, see Supplementary Table C (p. 184).

TABLE XXVI (see text, p. 74). LIMITS OF SIZE IN OTHER SPECIES

c = culmen, d = depth of beak, w = wing

	Geospiza					
	difficilis			*scandens* or *conirostris*		
	c	*d*	*w*	*c*	*d*	*w*
Culpepper	10·5–12·2	8·2– 9·5	71–77	13·6–16·2	12·3–19·1	77–
Wenman	10·0–11·7	7·7– 9·5	68–75	—	—	—
Tower	8·5–10·2	7·2– 8·5	60–66	13·0–16·4	11·1–15·3	74–
Abingdon	9·0–10·2	7·9– 9·0	61–64	13·7–15·9	8·9–10·7	72–
Bindloe	—	—	—	14·3–15·8	10·2–11·9	71–
James	9·5–11·4	8·5–10·0	65–76	12·0–13·8	8·2– 9·5	66–
Jervis	—	—	—	12·4–14·5	8·8– 9·7	69–
Indefatigable	9·1–10·2	8·2– 9·7	67–71	13·0–16·4	8·5–10·7	70–
Duncan	—	—	—	—	—	—
North Albemarle	—	—	—	—	—	—
South Albemarle	—	—	—	13·7–16·0	8·9–10·0	—
Barrington	—	—	—	12·8–16·4	9·3–11·4	68–
Chatham	—	—	—	11·7–14·0	9·3–10·2	—
Hood	—	—	—	13·0–17·4	13·3–18·7	74–
Charles	—	—	—	12·6–15·2	8·4–10·4	68–

	Camarhynchus						*Certhidea olivacea*	
	crassirostris			*pallidus*				
Island	*c*	*d*	*w*	*c*	*d*	*w*	*c*	
Culpepper	—	—	—	—	—	—	8·0–8·4	54
Wenman	—	—	—	—	—	—	7·4–8·2	5[illegible]
Tower	—	—	—	—	—	—	7·8–8·6	5[illegible]
Abingdon	9·8–11·0	11·3–13·0	—	—	—	—	7·6–8·4	50
Bindloe	—	—	—	—	—	—	7·8–8·9	5[illegible]
James	9·6–11·1	12·1–13·6	85–91	11·4–13·2	8·8–9·8	72–77	6·6–7·8	51
Jervis	—	—	—	—	—	—	—	
Indefatigable	9·9–11·0	11·5–13·1	84–89	11·3–12·7	8·8–9·5	70–74	7·0–8·1	5[illegible]
Duncan	—	—	—	—	—	—	7·0–8·4	51
North Albemarle	—	—	—	—	—	—	7·0–8·0	51
South Albemarle	9·6–10·9	10·6–12·9	79–88	10·4–12·2	7·4–9·0	68–77	6·9–8·0	51
Barrington	—	—	—	—	—	—	7·1–8·4	5[illegible]
Chatham	9·9–11·3	11·7–13·5	—	10·1–11·1	8·7–9·3	69–73	7·1–8·5	5[illegible]
Hood	—	—	—	—	—	—	7·4–8·6	5[illegible]
Charles	10·0–11·3	11·6–13·3	—	—	—	—	7·0–8·2	5[illegible]

	c	*d*	*w*
Camarhynchus heliobates Albemarle	9·5–11·4	7·6–8·8	68–75
Pinaroloxias inornata Cocos	9·0–11·2	5·5–7·0	64–71

Note. For number of specimens measured, see Supplementary Table C (p. 184).

TABLE XXVII (see text, p. 87). RATIO OF CULMEN (FROM NOSTRIL) TO WING IN *GEOSPIZA* SPECIES

Wing length	Mean culmen for birds of that wing length	Range of culmen for birds of that wing length	Average ratio of culmen : wing
	(i) *G. fuliginosa*		
60–62 (61·8)	8·6	7·7– 9·9	0·14
63	8·4	7·6– 9·8	0·13
64	8·5	7·4–10·2	0·13
65	8·6	7·7– 9·9	0·13
66	8·8	7·9–10·1	0·13
67–70 (67·7)	8·9	8·3– 9·5	0·13
	(ii) *G. fortis*		
66–68 (67·4)	10·9	10·1–11·8	0·16
69	11·4	10·4–12·3	0·17
70	11·4	10·7–12·1	0·16
71	11·6	10·1–13·9	0·16
72	11·6	10·7–13·2	0·16
73	11·7	10·3–13·5	0·16
74	12·3	11·3–13·4	0·17
75	12·4	10·8–14·0	0·17
76	12·8	12·1–13·6	0·17
77–80 (78)	12·9	11·9–14·1	0·17
	(iii) *G. magnirostris*		
77–80 (78·9)	15·2	14·0–16·5	0·19
81	15·4	14·2–16·8	0·19
82	15·4	14·0–16·0	0·19
83	15·7	14·5–17·3	0·19
84	16·1	14·7–17·7	0·19
85	16·4	15·9–17·2	0·19
86	16·0	15·1–17·2	0·19
87	15·8	14·2–16·8	0·18
88–90 (88·6)	16·4	15·1–17·4	0·19

Note. For *G. fuliginosa* and *G. fortis* all the individuals from Chatham, Charles and Albemarle were used. For *G. magnirostris* all males were taken. For number measured, see Supplementary Table D (p. 185).

TABLE XXVIII (see text, p. 87). RATIO OF DEPTH OF BEAK TO CULMEN IN *GEOSPIZA* SPECIES

Range of culmen	Mean culmen for this range	Mean depth	Range of depth	Ratio of depth : culmen
		(i) *G. fuliginosa*		
7·2– 7·7	7·5	7·7	7·1–8·3	1·0
7·8– 8·2	8·0	7·8	7·2–8·7	1·0
8·3– 8·7	8·5	8·1	7·2–8·9	1·0
8·8– 9·2	9·0	8·2	7·4–9·1	0·9
9·3– 9·7	9·5	8·4	7·4–9·1	0·9
9·8–10·2	9·9	8·3	8·0–8·6	0·8
		(ii) *G. fortis*		
9·8–10·2	10·1	11·0	10·1–11·9	1·1
10·3–10·7	10·5	11·5	10·7–12·1	1·1
10·8–11·2	11·0	11·9	10·2–13·2	1·1
11·3–11·7	11·4	12·3	10·8–13·6	1·1
11·8–12·2	12·0	12·8	11·1–15·3	1·1
12·3–12·7	12·5	13·4	12·0–14·8	1·1
12·8–13·2	12·9	14·4	12·1–16·0	1·1
13·3–13·7	13·5	14·3	13·1–15·4	1·1
13·8–14·2	14·0	14·9	13·6–16·0	1·1
		(iii) *G. magnirostris*		
13·8–14·2	14·0	17·3	15·8–19·4	1·2
14·3–14·7	14·5	18·0	16·3–19·4	1·2
14·8–15·2	15·0	18·9	16·8–21·6	1·3
15·3–15·7	15·5	19·6	18·0–21·4	1·3
15·8–16·2	16·0	20·2	18·2–22·3	1·3
16·3–16·7	16·4	20·9	18·9–22·5	1·3
16·8–17·2	17·0	20·9	19·5–22·9	1·2
17·3–17·7	17·4	21·2	19·9–23·2	1·2

Note. For number of specimens measured, see Supplementary Table D (p. 185).

TABLE XXIX (see text, pp. 91–92). RELATION OF VARIABILITY TO ABUNDANCE IN ISLAND FORMS OF THE SAME SPECIES

c = culmen, d = depth of beak, w = wing

Island	Size	Degree of isolation	Standard deviation of: Male			Female	
			c	d	w	c	d
		(i) *Geospiza magnirostris*					
Wenman	Extremely small	Extreme	0·64	0·89	—	—	—
Tower	Very small	Marked	0·60	1·13	1·4	0·59	0·62
Abingdon	Moderate	Moderate	0·66	0·94	1·9	0·63	0·78
Bindloe	Moderate	Moderate	0·77	0·95	2·1	0·70	0·97
James	Large	Very slight	0·70	1·03	2·1	1·00	1·31
Indefatigable	Large	Very slight	0·80	1·75	—	0·74	1·63
Jervis	Very small	Very slight	0·72	1·07	2·7	0·66	0·78

Note. At a rough estimate, there are several hundred *G. magnirostris* on Wenman, a very ew thousand on Tower, over ten thousand on other islands.

Island	Size	Degree of isolation	Male c	Male d	Male w	Female c	Female d
		(ii) *Geospiza fortis*					
Abingdon	Moderate	Moderate	0·51	0·66	2·1	0·35	0·54
Bindloe	Moderate	Moderate	0·47	0·41	1·3	0·32	0·47
Charles	Moderate	Moderate	0·88	1·22	2·5	0·82	0·96
Chatham	Large	Moderate	0·67	1·13	2·1	0·77	1·30
James	Large	Very slight	0·56	0·71	1·8	0·60	0·78
Indefatigable	Large	Very slight	0·81	1·32	2·8	0·68	1·18
N. Albemarle	Large	Very slight	0·73	1·06	2·1	0·88	1·21
S. Albemarle	Large	Very slight	0·83	1·17	2·3	0·99	1·42
Jervis	Very small	Very slight	0·38	—	—	—	—
Duncan	Very small	Very slight	0·73	0·86	2·3	0·55	0·95
Seymour	Very small	Very slight	0·78	0·96	1·7	0·66	0·92

Note. G. fortis is least variable where *G. magnirostris* is common, as on Abingdon, Bindloe, ames and Jervis. It is more variable where *G. magnirostris* is absent, as on Chatham and harles, or very scarce, as on Indefatigable and Albemarle.

land	Size	Degree of isolation	(iii) *Geospiza fuliginosa* Standard deviation of: Male c	Male w	Female c	(iv) *Certhidea olivacea* Standard deviation of: Male c	Male w	Female c
pper	Extremely small	Extreme	—	—	—	—	—	0·26
:	Very small	Marked	—	—	—	0·21	1·0	0·35
	Fairly small	Marked	0·36	1·5	0·43	0·28	1·1	0·37
ıgton	Fairly small	Small	0·37	1·6	0·24	0·30	0·9	0·31
don	Moderate	Moderate	0·39	2·3	0·35	0·23	0·7	0·27
e	Moderate	Moderate	0·39	1·8	0·28	0·31	0·9	0·32
s	Moderate	Moderate	0·40	1·6	0·49	0·29	1·0	0·31
am	Large	Moderate	0·68	1·5	0·67	0·34	0·9	0·38
	Large	Very slight	0·38	1·5	0·39	0·28	1·2	0·35
tigable	Large	Very slight	0·38	1·4	0·47	0·34	1·3	0·36
emarle	Large	Very slight	0·36	1·3	0·34	0·29	1·3	0·32
emarle	Large	Very slight	0·38	1·2	0·35	0·34	1·2	—
n	Very small	Very slight	0·41	1·5	0·43	0·47	1·4	—
ur	Very small	Very slight	0·36	1·2	0·35	—	—	—

. If under 10 specimens are available, no figure is given. For number of specimens measured, plementary Table C (p. 184, males) and Supplementary Table E (p. 185, females).

TABLE XXX (see text, p. 93). STANDARD DEVIATIONS FOR OTHER SPECIES (MALES)

c = culmen, d = depth, w = wing

	Geospiza						*Camarhynchus*											
	difficilis			*scandens* or *conirostris*			*crassirostris*			*psittacula*			*parvulus*			*pallidus*		
	c	*d*	*w*	*c*	*d*	*w*	*c*	*d*	*w*	*c*	*d*	*w*	*c*	*d*	*w*	*c*	*d*	*w*
Culpepper	0·43	0·31	1·4	0·87	2·17	2·9	—	—	—	—	—	—	—	—	—	—	—	—
Wenman	0·42	0·33	1·5	—	—	—	—	—	—	—	—	—	—	—	—	—	—	—
Tower	0·38	0·33	1·3	0·76	0·96	2·0	—	—	—	—	—	—	—	—	—	—	—	—
Abingdon	0·31	0·29	0·8	0·60	0·46	0·8	0·40	0·48	0·9	0·33	0·42	1·9	—	—	—	—	—	—
Bindloe	—	—	—	0·45	0·53	1·6	—	—	—	0·38	0·32	1·6	—	—	—	—	—	—
James	0·39	0·32	2·2	0·50	0·37	1·4	0·34	0·49	1·7	0·33	0·67	1·5	0·20	0·35	1·5	0·51	0·29	1·5
Jervis	—	—	—	0·63	0·30	1·4	—	—	—	—	—	—	—	—	—	—	—	—
Indefatigable	0·32	0·33	1·3	0·77	0·43	1·1	0·28	0·34	1·2	0·60	0·61	—	0·44	0·32	0·6	0·38	0·21	1·4
Duncan	—	—	—	—	—	—	—	—	—	—	—	—	—	—	—	—	—	—
North Albemarle	—	—	—	—	—	—	0·50	0·42	1·4	—	—	—	0·22	0·40	1·4	0·28	0·23	2·0
South Albemarle	—	—	—	0·71	0·37	1·1	0·28	0·53	1·9	0·42	0·48	1·9	0·32	0·33	1·1	0·43	0·40	1·5
Barrington	—	—	—	0·70	0·48	1·3	—	—	—	—	—	—	—	—	—	—	—	—
Chatham	—	—	—	0·91	0·37	—	0·32	0·44	1·5	—	—	—	0·31	0·33	1·7	0·30	0·23	1·7
Charles	—	—	—	0·62	0·39	1·5	0·35	0·48	1·3	—	—	—	0·30	0·28	1·5	—	—	—

	c	*d*	*w*
G. conirostris Hood	0·94	1·10	2·4
G. conirostris Gardner near Hood	0·99	1·11	2·4
C. pauper Charles	0·40	0·43	1·5
C. heliobates Albemarle	0·47	0·28	1·6
P. inornata Cocos	0·38	0·25	1·3

Note. For number of specimens measured, see Supplementary Table C (p. 184).

TABLE XXXI (see text, pp. 93–94). VARIABILITY OF DIFFERENT SPECIES

Species	Estimate of relative abundance expressed as percentage of all individuals of all species present	Coefficient of variability of		
		Culmen	Depth of beak	Wing
	(i) On Tower			
Certhidea olivacea	44	2·5	—	1·8
Geospiza difficilis	44	4·0	4·2	2·1
G. magnirostris	9	3·6	5·3	1·6
G. conirostris	3	5·3	7·4	2·6
	(ii) On Indefatigable			
Certhidea olivacea	47	4·5	—	2·4
Geospiza fuliginosa	16	4·5	4·7	2·2
G. scandens	12	5·1	4·4	1·5
G. fortis	10	6·8	10·3	3·8
Camarhynchus parvulus	5	5·9	4·4	—
C. crassirostris	3	2·7	2·8	1·4
C. pallidus	3	3·1	2·3	1·9
Geospiza magnirostris	2	5·2	9·2	—
G. difficilis	1	3·3	3·8	1·8
Camarhynchus psittacula	1	6·2	5·7	—
	(iii) Some Other Species			
Geospiza fuliginosa Chatham	—	7·7	5·4	2·4
G. conirostris Culpepper	—	5·8	7·4	3·5
G. conirostris Hood	—	6·1	6·9	3·0
Camarhynchus pauper Charles	—	4·5	4·9	2·2
C. pallidus South Albemarle	—	3·8	4·9	2·1
C. heliobates Albemarle	—	4·5	—	2·2
Pinaroloxias inornata Cocos	—	3·7	4·1	1·8
Passer domesticus England	—	4·0	4·0	2·4
11 forms of *Lanius ludovicianus*, North America	—	2·5–5·0	2·2–2·9	1·1–1·9
24 forms of *Junco* species, North America	—	3·4–5·3	3·5–6·1	2·1–3·2

Notes. (i) For standard deviations, see Tables XXIX and XXX, and for number measured, Supplementary Table C (p. 184).

(ii) The estimates of relative abundance were made during residence on Indefatigable and during a short visit to Tower, and though rough, they are sufficiently accurate to show that there is no obvious relation between variability and abundance.

TABLE XXXII (see text, p. 98). INTERMEDIATE AND FREAK SPECIMENS

Intermediate between	Island	No.	Sex	Culmen	Depth of beak	Wing	Formerly named
(i) In *Geospiza*							
G. magnirostris and *G. fortis*	James	1	Male	13·8	16·1	81	'*G. bauri*'
,,	James	3	Female	13·7–14·0	15·3–15·5	74–78	,,
,,	Bindloe	1	Female	13·4	15·4	73	,,
,,	Bindloe	1	?	13·0	15·7	74	,,
G. fortis and *G. fuliginosa*	Chatham	1	Male	9·6	10·2	69	'*G. harterti*'
,,	Hood	1	Male	8·8	*c.* 10·2	65	,,
G. fortis and *G. d. debilirostris*	James	1	Male	11·4	10·0	69	—
G. fortis and *G. scandens*	Charles	2	Male	13·0–13·2	11·0–11·6	69–73	'*Cactornis brevirostris*'
,,	Indefatigable	2	?	12·8	11·0–13·0	68–71	,,
(ii) In *Camarhynchus*							
C. psittacula and *C. parvulus*	South Albemarle	2	Male	7·4–8·4	8·6–8·7	63	—
,,	South Albemarle	4	Female	7·4–7·9	7·3–8·0	63–64	—
,,	South Albemarle	2	?	7·6–7·8	7·9–8·0	64	—
C. pauper and *C. parvulus*	Charles	2	Male	7·8–8·6	7·6–8·7	65–66	—
(iii) Intergeneric							
Camarhynchus and *Certhidea*	Charles	2	Male	7·8–7·9	5·4–5·9	59–60	'*Camarhynchus conjunctus*'
,,	Chatham	1	Male	8·0	*c.* 6·2	59	'*Camarhynchus aureus*'
(iv) Dwarf specimens							
Camarhynchus crassirostris	Narborough	1	Male	8·6	9·8	77	
Camarhynchus	Indefatigable	1	Male	10·4	6·4	65	'*Cactospiza*

Supplementary Table A

Number of Specimens available for Tables XVIII and XIX

(see p. 168)

	Camarhynchus			*Geospiza*				
land	*crassi-rostris*	*psitta-cula*	*par-vulus*	*magni-rostris*	*fortis*	*fuligi-nosa*	*diffi-cilis*	*scandens* or *coni-rostris*
pper	—	—	—	—	—	—	33	—
ıan	—	—	—	—	—	—	67	—
r	—	—	—	29	—	—	53	43
'don	16	18	—	51	48	48	19	—
oe	—	24	—	41	36	24	—	—
s	26	17	28	46	48	23	44	25
	—	—	—	35	—	—	—	—
atigable	28	—	19	—	107	81	33	85
an	—	—	—	—	30	84	—	—
arle	67	17	25	—	135	152	—	—
ngton	—	—	—	—	—	41	—	61
am	37	—	91	—	110	127	—	—
and dner	—	—	—	—	—	40	—	159
s	25	80[1]	87	—	182	86	—	104

[1] *C. pauper.*

Supplementary Table B

Number of Specimens measured in Tables XX and XXI

(pp. 169, 170).

	Males			
Species	Fully black	Partly black	Without black	Females
G. magnirostris	16	12	18	21
G. fortis	56	25	28	64
G. fuliginosa	62	34	28	108
		Partly black and without black		
G. difficilis	26	41		35
G. scandens	40	44		32
G. conirostris	38	32		42
C. crassirostris	—	34	23	26
C. psittacula	—	13	11	14
C. parvulus	—	60	26	38
P. inornata	78	17	21	43
		All males		
C. pallidus	—	55		24
C. olivacea	—	43		39

Supplementary Table C

Number of Males measured for Tables XI–XIII and XXII–XXXI (omitting XXVII and XXVIII)

Note. The first figure for each form is the number of beaks and the second figure the number of wings measured.

	Geospiza					*Camarhynchus*				
Island	*magni-rostris*	*fortis*	*fuligi-nosa*	*difficilis*	*scandens* or *coni-rostris*	*crassi-rostris*	*psitta-cula*	*parvulus*	*pallidus*	*Certhidea olivacea*
Culpepper	—	—	—	32 17	17 17	—	—	—	—	6 6
Wenman	10 9	—	—	67 26	—	—	—	—	—	9 10
Tower	29 20	—	—	52 43	43 30	—	—	—	—	41 41
Abingdon	50 18	48 11	48 13	18 11	17 4	16 8	18 8	—	—	17 18
Bindloe	41 14	36 14	24 11	—	9 5	5 —	24 13	—	—	18 18
James	46 16	46 24	23 18	44 32	23 15	26 20	17 14	28 18	13 13	31 31
Jervis	35 22	10 —	12 —	—	15 11	—	—	—	—	—
Indefatigable	13 4	105 31	80 32	33 23	87 40	28 16	7 —	19 5	9 8	44 43
Duncan	—	30 10	84 45	—	—	—	—	—	—	21 21
North Albemarle	—	65 30	87 60	—	—	10 6	—	14 6	6 6	20 19
South Albemarle	8 —	70 27	63 37	—	16 5	56 34	17 7	24 16	53 55	28 28
Barrington	—	—	41 26	—	61 37	—	—	—	—	40 38
Chatham	—	109 56	125 62	—	7 —	37 9	—	91 4	8 8	46 46
Hood	—	—	32 24	—	87 64	—	—	—	—	44 45
Charles	—	181 102	86 41	—	102 59	24 5	3 —	86 60	—	25 25
Darwin's specimens	3 3	—	—	2 2	—	—	—	—	—	—

G. fortis, Daphne	31	23
G. fuliginosa, Crossman	12	8
G. conirostris, Gardner near Hood	72	38
C. pauper, Charles	80	21
C. heliobates, Albemarle	32	32
Pinaroloxias inornata, Cocos	124	78

Supplementary Table D

Number of Specimens measured in Tables XXVII and XXVIII (pp. 177–178).

For Table XXVII

G. fuliginosa		*G. fortis*		*G. magnirostris*	
Wing length	No. measured	Wing length	No. measured	Wing length	No. measured
60–62	29	66–68	9	77–80	9
63	52	69	9	81	13
64	68	70	27	82	13
65	44	71	44	83	17
66	26	72	38	84	13
67–70	6	73	28	85	13
—	—	74	17	86	14
—	—	75	26	87	10
—	—	76	12	88–90	7
—	—	77–80	18	—	—

For Table XXVIII

G. fuliginosa		*G. fortis*		*G. magnirostris*	
Culmen length	No. measured	Culmen length	No. measured	Culmen length	No. measured
7·5	13	10·1	5	14·0	11
8·0	28	10·5	12	14·5	13
8·5	51	11·0	54	15·0	43
9·0	65	11·4	55	15·5	40
9·5	16	12·0	72	16·0	55
9·9	9	12·5	25	16·4	32
—	—	12·9	33	17·0	19
—	—	13·5	10	17·4	7
—	—	14·0	6	—	—

Supplementary Table E

Number of Female Specimens measured in Table XXIX (p. 179.)

Island	*G. magnirostris*	*G. fortis*	*G. fuliginosa*	*C. olivacea*
Culpepper	—	—	—	13
Tower	17	—	—	21
Hood	—	—	15	19
Barrington	—	—	10	33
Abingdon	29	23	26	14
Bindloe	11	17	21	19
Charles	—	93	87	22
Chatham	—	64	108	23
James	21	23	12	24
Indefatigable	13	49	43	38
North Albemarle	—	34	57	11
South Albemarle	—	56	39	—
Jervis	13	—	—	—
Duncan	—	20	30	—
Seymour	—	31	25	—

REFERENCES

We are as dwarfs mounted on the shoulders of giants, so that we can see more and further than they; yet not by virtue of the keenness of our eyesight, nor through the tallness of our stature, but because we are raised and borne aloft on that giant mass.

BERNARD OF CHARTRES, twelfth century A.D.

REFERENCES have been cited in the standard way. In the text is given the name of the author followed by the year of publication, while the full references are listed below under authors alphabetically. This bibliography includes only those works mentioned in the text, and where a general work provides detailed references to a topic here touched on only briefly, reference in this book is usually given only to the general work, which should be consulted for the detailed references. In particular, the species problem has been treated generally by Dobzhansky (1937, 1941) from the genetical standpoint, by Mayr (1942) from the systematic standpoint, and by Huxley (1940, 1942), who presents a general synthesis, while further references to the ecology of closely related bird species are given in Lack (1944*a*). A complete bibliography of Galapagos birds is provided by Swarth (1931).

A preliminary summary of the present work on Darwin's finches appeared in 1940 (Lack, 1940*b*), while a detailed account was completed in May 1940 for publication by the California Academy of Sciences, but has not yet appeared at the time of writing. The latter paper includes more details on breeding habits, plumage and measurements than are given here, while in certain respects my views have changed since it was written, notably in regard to the adaptive significance of the beak differences between species and island forms, discussed particularly in Chapters VI and XV of this book. As regards these and other differences between the two accounts, the present book represents a more mature opinion. Observations carried out at the California Academy of Sciences by Dr Robert T. Orr, on the captive finches which we brought home from the Galapagos, have also not appeared at the time of writing.

DETAILED REFERENCES

BANKS, N. (1902). Papers from the Hopkins-Stanford Galapagos Expedition, 1898–9. VII. Entomological results (6). Arachnida. *Proc. Wash. Acad. Sci.* **4**, 49–86.

BANNERMAN, D. (Unpublished MS.) The History of the Birds of the Canary Islands (deposited in British Museum of Natural History).

BARLOW, N. (1933). *Charles Darwin's Diary of the Voyage of H.M.S. 'Beagle'*, pp. xiii, 333–43.

BARLOW, N. (1935). Charles Darwin and the Galapagos Islands. *Nature* **136**, 391.

BARRETT-HAMILTON, G. E. H. and HINTON, M. A. C. (1913–14). *A History of British Mammals*, **2**, 249–348, 405–66.

BATES, G. L. (1931). On geographical variation within the limits of West Africa: some generalizations. *Ibis*, pp. 255–302.

BAUR, G. (1891). On the origin of the Galapagos Islands. *Amer. Nat.* **25**, 217–29, 307–19, 902–7.

BAUR, G. (1897). On the origin of the Galapagos Islands. *Amer. Nat.* **31**, 661–80, 864–96.

BEAUCHAMP, R. and ULLYOTT, P. (1932). Competitive relationships between certain species of fresh-water triclads. *J. Ecol.* **20**, 200–8.

BEEBE, W. (1924). *Galapagos: World's End.*

BEHLE, W. H. (1942). Distribution and variation of the horned larks (*Otocoris alpestris*) of western North America. *Univ. Calif. Publ. Zool.* **46**, 205–316.

BOYD, A. W. (1936). Report on the swallow enquiry, 1935. *Brit. Birds*, **30**, 98–116.

BULLER, W. L. (1888). A History of the Birds of New Zealand. 2nd ed. **1**, pp. 7, 10.

BUXTON, P. A. (1938). The formation of species among insects in Samoa and other oceanic islands. *Proc. Linn. Soc. Lond.* **150**, 264–7.

CHAPMAN, F. M. (1926). The distribution of bird life in Ecuador. *Bull. Amer. Mus. Nat. Hist.* **55**, 594.

CLARKE, W. E. (1905). Ornithological results of the Scottish National Antarctic Expedition. I. On the birds of Gough Island, South Atlantic Ocean. *Ibis*, pp. 255–8.

COLNETT, J. (1798). *A Voyage to the South Atlantic and round Cape Horn into the Pacific Ocean.*

COWLEY, A. (1699). *Voyage round the World.*

DALL, W. H. and OSCHNER, W. H. (1928*a*). Tertiary and Pleistocene Mollusca from the Galapagos Islands. *Proc. Calif. Acad. Sci.* (4), **17**, 89–139.

DALL, W. H. and OSCHNER, W. H. (1928*b*). Land shells of the Galapagos islands. *Proc. Calif. Acad. Sci.* (4), **17**, 141–84.

DAMPIER, W. (1717). *A New Voyage round the World.* Ed. J. Masefield (1906).

DARWIN, C. (1839). *Journal of Researches into the Geology and Natural History of the various countries visited during the Voyage of H.M.S. 'Beagle'*, etc. (Quotations are made from the revised 2nd edition, which first appeared in 1845, and has been reprinted several times under modified titles.)

DARWIN, C. (1841). *The Zoology of the Voyage of H.M.S. 'Beagle'*. Pt. III. Birds, by J. Gould. Notice of their habits and ranges, by Charles Darwin, pp. 98–106.

DARWIN, C. (1859). *On the Origin of Species by means of Natural Selection.* (Quotations are made from the second edition.)

DARWIN, F. (1887). *Letters of Charles Darwin* (edited), 3rd ed. **2**, 23.

DAY, B. and FISHER, R. A. (1937). The comparison of variability in populations having unequal means. *Ann. Eugen., Lond.*, **7**, esp. pp. 337–40.

DOBZHANSKY, TH. (1937). *Genetics and the Origin of Species.* Also 2nd ed. 1941.

FISHER, R. A. (1937). The relation between variability and abundance shown by the measurements of the eggs of British nesting birds. *Proc. Roy. Soc.* B, **122**, 1–26.

FitzRoy, Capt. R. (1839). *Narrative of the surveying voyages of His Majesty's ships 'Adventure' and 'Beagle'*, **2**, 484–505.

Gause, G. F. (1934). *The Struggle for Existence*, esp. pp. 19–20.

Gifford, E. W. (1919). Field notes on the land birds of the Galapagos Islands and of Cocos Island, Costa Rica. *Proc. Calif. Acad. Sci.* (4), **2**, 189–258.

Gilbert, P. A. (1939). The bower-painting habit of the satin bower-bird (*Ptilonorhynchus violaceus*). *Emu*, **39**, 18–22.

Gould, J. (1837). Description of new species of finches collected by Darwin in the Galapagos. *Proc. Zool. Soc. Lond.* **5**, 4–7.

Gould, J. (1841). In Darwin, C. (1841). *The Zoology of the Voyage of H.M.S. 'Beagle'*. Pt. III. Birds, pp. 98–106.

Gould, J. (1843). Descriptions of nine new species of birds collected during the recent voyage of H.M.S. *Sulphur*. *Proc. Zool. Soc. Lond.* pp. 103–6. (Describes *Pinaroloxias inornata*.)

Gould, J. (1844). *The Zoology of the Voyage of H.M.S. 'Sulphur'*. Ed. R. B. Hinds.

Griscom, L. (1937). A monographic study of the red crossbill (*Loxia curvirostra*). *Proc. Boston Nat. Hist. Soc.* **41**, 75–209.

Gulick, A. (1932). Biological peculiarities of oceanic islands. *Quart. Rev. Biol.* **7**, 405.

Hagen, Y. (1940). In Christopherson, E. *Tristan da Cunha. The Lonely Isle.* Eng. trs. R. L. Benham, pp. 96–99.

Hawkins, Sir R. (1593). *Observations in a Voiage into the South Seas.*

Hebard, M. (1920). Expedition of the California Academy of Sciences to the Galapagos Islands, 1905–6. XVII. Dermaptera and Orthoptera. *Proc. Calif. Acad. Sci.* (4) **2**, 311–46.

Hellmayr, C. E. (1927). Catalogue of Birds of the Americas and adjacent islands in the Field Museum of Natural History. Pt. V. Tyrannidae. *Field. Mus. Nat. Hist. Publ.* no. 242, Zool. Ser. **13**, 93–4.

Hellmayr, C. E. (1938). Catalogue of Birds of the Americas and adjacent islands in the Field Museum of Natural History. Pt. XI. Ploceidae-Catamblyrhynchidae-Fringillidae, pp. 130–46.

Hillebrand, W. (1888). *Flora of the Hawaiian Islands.*

Hindwood, K. (1940). The birds of Lord Howe Island. *Emu*, **40**, 71–2.

Hooker, J. D. (1847). On the vegetation of the Galapagos archipelago, as compared with that of some other tropical islands and of the continent of America. *Trans. Linn. Soc. Lond. Bot.* **20**, 235–62.

Howell, J. T. (1941). The Templeton Crocker Expedition of the California Academy of Sciences, 1932. No. 40. The genus *Scalesia*. *Proc. Calif. Acad. Sci.* (4), **22**, 221–71.

Huxley, J. S. (1927). *Problems of Relative Growth.*

Huxley, J. S. (1942). *Evolution: the Modern Synthesis.*

Huxley, J. S. (1943). Evolution in action. (Review of Mayr, 1942.) *Nature, Lond.* **151**, 347–8.

Huxley, J. S. *et al.* (1940). *The New Systematics.*

Lack, D. (1933). Habitat selection in birds. *J. Anim. Ecol.* **2**, 239–62. See also (1937), *Brit. Birds*, **31**, 130–6 and (1940), *Brit. Birds*, **34**, 80–4.

Lack, D. (1940*a*). Variation in the introduced English sparrow. *Condor*, **42**, 239–41.

Lack, D. (1940*b*). Evolution of the Galapagos finches. *Nature, Lond.*, **146**, 324–7.

Lack, D. (1940*c*). The Galapagos finches. *Bull. Brit. Orn. Cl.* **60**, 46–50.

Lack, D. (1944*a*). Ecological aspects of species formation in passerine birds. *Ibis*, **86**, 260–86.

Lack, D. (1944*b*). Correlation between beak and food in the crossbill *Loxia curvirostra*. *Ibis*, **86**, 552–3.

Lack, D. (1945). The Galapagos finches (Geospizinae): a study in variation. *Occ. Pap. Calif. Acad. Sci.* **21**, 1–159.

Linell, M. (1898). On the coleopterous insects of the Galapagos Islands. *Proc. U.S. Nat. Mus.* **21**, 249–68.

Lowe, P. R. (1923). Notes on some land birds of the Tristan da Cunha group collected by the *Quest* expedition. *Ibis*, pp. 519–23.

Lowe, P. R. (1930). Hybridisation in birds in its possible relation to the evolution of the species. *Bull. Brit. Orn. Cl.* **50**, 22–9.

Lowe, P. R. (1936). The finches of the Galapagos in relation to Darwin's conception of species. *Ibis*, pp. 310–21.

Lynes, H. (1930). Review of the genus *Cisticola*. *Ibis*, suppl. no., pp. 1–673.

Mathews, G. M. (1928). *The Birds of Norfolk and Lord Howe Islands*, pp. 50–3.

Mayne, B. and Young, M. D. (1938). Antagonism between species of malaria parasites in induced mixed infections. Preliminary note. *U.S. Publ. Hlth Rep.* **53**, no. 30, pp. 1289–91.

Mayr, E. (1931). A systematic list of the birds of Rennell Island with descriptions of new species and subspecies. *Amer. Mus. Novit.* no. 486, pp. 18, 26–7.

Mayr, E. (1932*a*). Notes on Meliphagidae from Polynesia and the Solomon Islands. *Amer. Mus. Novit.* no. 516, pp. 5–9.

Mayr, E. (1932*b*). Notes on thickheads (*Pachycephala*) from the Solomon Islands. *Amer. Mus. Novit.* no. 522. Also next.

Mayr, E. (1932*b*). Notes on thickheads (*Pachycephala*) from Polynesia. *Amer. Mus. Novit.* no. 531.

Mayr, E. (1933*a*). Notes on Polynesian flycatchers and a revision of the genus *Clytorhynchus* Elliot. *Amer. Mus. Novit.* no. 628, pp. 17–20.

Mayr, E. (1933*b*). Notes on the genera *Myiagra* and *Mayrornis*. *Amer. Mus. Novit.* no. 651.

Mayr, E. (1933*c*). Notes on the variation of immature and adult plumages in birds and a physiological explanation of abnormal plumages. *Amer. Mus. Novit.* no. 666.

Mayr, E. (1934). Notes on the genus *Petroica*. *Amer. Mus. Novit.* no. 714.

Mayr, E. (1938). Notes on New Guinea birds. IV. *Amer. Mus. Novit.* no. 1006, pp. 10–11.

Mayr, E. (1942). *Systematics and the Origin of Species*.

McNeill, J. (1901). Papers from the Hopkins-Stanford Galapagos Expedition, 1898–1899. IV. Entomological results (4). Orthoptera. *Proc. Wash. Acad. Sci.* **3**, 487–506.

Meise, W. (1936*a*). Ueber Artentstehung durch Kreuzung in der Vogelwelt. *Biol. Zbl.* **56**, 590–604.

Meise, W. (1936*b*). Zur Systematik und Verbreitungsgeschichte der Haus- und Weidensperlinge *Passer domesticus* (L.) und *hispaniolensis* (T.). *J. Orn. Lpz.* **84**, 631–72.

Melliss, J. C. (1875). *St Helena*, pp. 283–6.

Miller, A. H. (1931). Systematic revision and natural history of the American shrikes (*Lanius*). *Univ. Calif. Publ. Zool.* **38**, 11–242.

Miller, A. H. (1941). Speciation in the avian genus *Junco*. *Univ. Calif. Publ. Zool.* **44**, 173–434.

Muller, H. J. (1940). Bearings of the *Drosophila* work on systematics. *The New Systematics* (ed. J. S. Huxley), pp. 185–268, esp. pp. 194–8.

Murphy, R. C. (1938*a*). The need of insular exploration as illustrated by birds. *Science*, **88**, 533–9.

Murphy, R. C. (1938*b*). On pan-antarctic terns. *Amer. Mus. Novit.* no. 977, pp. 5–6.

Murphy, R. C. and Chapin, J. P. (1929). A collection of birds from the Azores. *Amer. Mus. Novit.* no. 384, pp. 20–1.

Niethammer, G. (1937). *Handbuch der deutschen Vogelkunde*, **1**, 81–8, 413–20.

Niethammer, G. (1940). Die Schutzanpassung der Lerchen. From Hoesch, W. and Niethammer, G., *Die Vogelwelt Deutsch-Südwestafrikas*. *J. Orn.* **88**, 75–83. (Quoted by Mayr, 1942, p. 86.)

ORR, R. T. (1945). A study of captive Galapagos finches of the genus *Geospiza*. *Condor*, **47**, 177–205.

PERKINS, R. C. L. (1903). *Fauna Hawaiiensis*, vol. **1**, pt. IV, Vertebrata: Aves, pp. 368–466.

PERKINS, R. C. L. (1913). *Fauna Hawaiiensis*, **1**, pt. VI, Introduction, being a review of the land-fauna of Hawaii, pp. xv–ccxxviii.

PORTER, D. (1822). *Journal of a Cruise made to the Pacific*, etc. (2nd ed.).

RENSCH, B. (1933). Zoologische Systematik und Artbildungsproblem. *Verh. dtsch. Zool. Ges.* **35** (*Zool. Anz.* suppl. 6), pp. 19–83, esp. pp. 37–8.

RIDGWAY, R. (1897). Birds of the Galapagos archipelago. *Proc. U.S. Nat. Mus.* **19**, no. 1116, pp. 459–670.

ROBINSON, B. L. (1902). Flora of the Galapagos Islands. *Proc. Amer. Acad. Arts Sci.* **38**; contr. *Gray Herb.* **24**, 77–269.

ROBSON, G. C. and RICHARDS, O. W. (1936). *The Variation of Animals in Nature*, esp. pp. 276–9.

ROGERS, WOODES (1712). *A Cruising Voyage round the World.*

ROSE, R. (1924). Man and the Galapagos. In Beebe, W. *Galapagos: World's End*, pp. 332–417.

ROTHSCHILD, W. (1893–1900). *The Avifauna of Laysan and the neighbouring Islands, with a complete History to date of the Birds of the Hawaiian Possessions.*

ROTHSCHILD, W. and HARTERT, E. (1899). A review of the ornithology of the Galapagos Islands. *Novit. Zool.* **6**, 85–205.

ROTHSCHILD, W. and HARTERT, E. (1902). Further notes on the fauna of the Galapagos Islands. Notes on the birds. *Novit. Zool.* **9**, 381–418.

SALVIN, O. (1876). On the avifauna of the Galapagos archipelago. *Trans. Zool. Soc. Lond.* **9**, 447–510.

SCHARFF, R. (1912). *Distribution and Origin of Life in America*, ch. XII The Galapagos Islands, pp. 295–335.

SCLATER, P. L. and SALVIN, O. (1870). Characters of new species of birds collected by Dr Habel in the Galapagos Islands. *Proc. Zool. Soc. Lond.* pp. 322–7.

SIMPSON, G. G. (1943). Turtles and the origin of the fauna of Latin America. *Amer. J. Sci.* pp. 413–29, esp. p. 420.

SMITH, E. A. (1892). On the land-shells of St Helena. *Proc. Zool. Soc. Lond.* pp. 258–70.

SNODGRASS, R. E. (1902*a*). The relation of the food to the size and shape of the bill in the Galapagos genus *Geospiza*. *Auk*, **19**, 367–81.

SNODGRASS, R. E. (1902*b*). *Schistocera*, *Sphingonotus* and *Halmenus*. Papers from the Hopkins-Stanford Galapagos Expedition, 1898–1899. VIII. Entomological results (7). *Proc. Wash. Acad. Sci.* **4**, 411–55.

SNODGRASS, R. E. (1903). Notes on the anatomy of *Geospiza*, *Cocornis* and *Certhidea*. *Auk*, **20**, 402–17.

SNODGRASS, R. E. and HELLER, E. (1904). Papers from the Hopkins-Stanford Galapagos Expedition, 1898–99. XVI. Birds. *Proc. Wash. Acad. Sci.* **5**, 231–372.

STEWART, A. (1911). Expedition of the California Academy of Sciences to the Galapagos Islands, 1905–1906. II. A botanical survey of the Galapagos Islands. *Proc. Calif. Acad. Sci.* (4), **1**, 7–288.

STEWART, A. (1915). Some observations concerning the botanical conditions on the Galapagos islands. *Trans. Wis. Acad. Sci. Arts Lett.* **18**, 272–339.

STRESEMANN, E. (1931). Die Zosteropiden der indo-australischen Subregion. *Mitt. Zool. Mus. Berl.* **17**, 201–38.

STRESEMANN, E. (1936). Zur Frage der Artbildung in der Gattung *Geospiza*. *Orgaan der Club Van Nederlandische Vogelkundigen*, **9**, no. 1, pp. 13–21.

STRESEMANN, E. (1939). *Zosterops siamensis* Blyth—eine gelbbaüchige Rasse von *Zosterops palpebrosa*. *J. Orn., Lpz.*, **87**, 156–64.

SUSHKIN, P. P. (1925). The evening grosbeak (*Hesperiphona*), the only American genus of a Palaearctic group. *Auk*, **42**, 256–61.

SUSHKIN, P. P. (1929). On some peculiarities of adaptive radiation presented by insular faunae. *Verh. VI. Int. Orn. Kong.* 1926, pp. 375–8.

SWARTH, H. S. (1929). A new bird family (Geospizidae) from the Galapagos Islands. *Proc. Calif. Acad. Sci.* (4), **18**, 29–43.

SWARTH, H. S. (1931). The avifauna of the Galapagos Islands. *Occ. Pap. Calif. Acad. Sci.* **18**.

SWARTH, H. S. (1934). The bird fauna of the Galapagos Islands in relation to species formation. *Biol. Rev.* **9**, 213–34.

SWARTH, H. S. (1935). Injury-feigning in nesting birds. *Auk*, **52**, 353.

TEN KATE, C. G. B. *et al.* (1937). *De Nederlandsche Vogels*, pp. 83, 85, 87.

TICEHURST, C. B. (1938). *A Systematic Review of the genus* Phylloscopus (Brit. Mus. Publ.).

USINGER, R. L. (1941). Problems of insect speciation in the Hawaiian Islands. *Amer. Nat.* **75**, 251–63.

VAN DENBURGH, J. (1912*a*). Expedition of the California Academy of Sciences to the Galapagos Islands, 1905–1906. IV. The snakes of the Galapagos Islands. *Proc. Calif. Acad. Sci.* (4), **1**, 323–74.

VAN DENBURGH, J. (1912*b*). Expedition of the California Academy of Sciences to the Galapagos Islands, 1905–1906. VI. The geckos of the Galapagos archipelago. *Proc. Calif. Acad. Sci.* (4), **1**, 405–30.

VAN DENBURGH, J. (1914). Expedition of the California Academy of Sciences to the Galapagos Islands, 1905–1906. X. The gigantic land tortoises of the Galapagos archipelago. *Proc. Calif. Acad. Sci.* (4), **2**, 203–374.

VAN DENBURGH, J. and SLEVIN, J. R. (1913). Expedition of the California Academy of Sciences to the Galapagos Islands, 1905–1906. IX. The Galapagoan lizards of the genus *Tropidurus*; with notes on the iguanas of the genera *Conolophus* and *Amblyrhynchus*. *Proc. Calif. Acad. Sci.* (4), **2**, 133–202.

WAFER, L. (1699). *The New Voyage and Description of the Isthmus of America.* Rep. by Hakluyt Society (1933).

WALLACE, A. R. (1880). *Island Life.*

WHEELER, W. M. (1919). Expedition of the California Academy of Sciences to the Galapagos Islands, 1905–1906. The ants of the Galapagos Islands. *Proc. Calif. Acad. Sci.* (4), **2**, 259–97.

WILKINS, G. H. (1923). Report on the birds collected during the 'Quest' (Shackleton-Rowett Expedition) to the South Antarctic. *Ibis*, pp. 496–7, 505–6.

WILLIAMS, F. X. (1911). Expedition of the California Academy of Sciences to the Galapagos Islands, 1905–1906. The butterflies and hawk-moths of the Galapagos Islands. *Proc. Calif. Acad. Sci.* (4), **1**, 289–322.

WILLIAMS, F. X. (1926). Expedition of the California Academy of Sciences to the Galapagos Islands, 1905–1906. The bees and aculeate wasps of the Galapagos Islands. *Proc. Calif. Acad. Sci.* (4), **2**, 347–57.

WILSON, E. A. (1907). *Mammalia* (*Whales and Seals*). *Nat. Antarctic Exped.* 1901–4. *Nat. Hist.* II, 10–50, esp. pp. 33–46. Brit. Mus. Publ.

WITHERBY, H. F. *et al.* (1938). *The Handbook of British Birds*, **1**.

WITHERBY, H. F. and FITTER, R. S. R. (1942). Black redstarts in England in the summer of 1942. *Brit. Birds*, **36**, 132–9.

WOLLASTON, T. V. (1877). *Coleoptera Sanctae Helenae.*

WORTHINGTON, E. B. (1940). Geographical differentiation in fresh waters with special reference to fish. *The New Systematics*, ed. J. S. Huxley, pp. 287–302.

WRIGHT, SEWALL (1940). Breeding structure of populations in relation to speciation. *Amer. Nat.* **74**, 232–48. Also next.

WRIGHT, SEWALL (1940). The statistical consequences of Mendelian heredity in relation to speciation. *The New Systematics*, ed. J. S. Huxley, pp. 161–83.

INDEX OF ANIMALS AND PLANTS

Note. Scientific Latin names are listed alphabetically under both genus and species, but the page reference is given only under the genus. English names are listed only under the group name; e.g. the tree pipit *Anthus trivialis* is listed under both *Anthus* and *trivialis*, also under pipit, but not under tree. When, in this index, a scientific Latin name follows the English name, page references are given only after the Latin name. The figures in heavy type refer to illustrations.

INDEX OF PERSONS, PLACES AND INSTITUTIONS

INDEX OF SUBJECT-MATTER

Note. A further guide to subject-matter is provided by the table of contents, pp. vii–viii.